Markham Clements R.

Major James Rennell and Modern English Geography

Markham Clements R.

Major James Rennell and Modern English Geography

ISBN/EAN: 9783337321093

Printed in Europe, USA, Canada, Australia, Japan

Cover: Foto ©berggeist007 / pixelio.de

More available books at **www.hansebooks.com**

THE CENTURY SCIENCE SERIES.

MAJOR JAMES RENNELL

AND THE RISE OF

MODERN ENGLISH GEOGRAPHY

BY

CLEMENTS R. MARKHAM, C.B., F.R.S.

President of the Royal Geographical Society and
President of the Hakluyt Society

New York

MACMILLAN & CO.

1895

PREFACE.

JAMES RENNELL was the greatest geographer that Great Britain has yet produced. His pre-eminence, as Sir Henry Yule said in 1881, is still undisputed. But this is not the sole reason for selecting him as the representative of geography. He was not only the greatest, he was also the most many-sided devotee of the science. He was an explorer both by sea and land, a map compiler, a physical geographer, a critical and comparative geographer, and a hydrographer.

When the present writer had occasion to prepare a notice of Major Rennell for his "Memoir of the Indian Surveys," * he had some difficulty in finding materials. He consulted Sir Henry Yule in 1878, who took great interest on general grounds, but especially because Rennell was the most distinguished ornament of the corps to which he himself belonged —the Bengal Engineers. The interest was renewed by the discovery of a porcelain medallion of Major Rennell at the India Office, by Sir George Birdwood. Sir Henry had it photographed, and the result was so successful that the editor of the *Royal Engineers' Journal* proposed to publish it, and requested Sir Henry to furnish a sketch of Rennell's career as an accompaniment. With his usual industry, Sir Henry

* 1871. Second edition, 1878, p. 54.

Yule set to work to collect materials, and was so successful that he was able to contribute a very brief, but exceptionally valuable and interesting, memoir of sixteen pages to the *Royal Engineers' Journal* in 1881. He entertained a hope of utilising the materials he had brought together at some future time in a memoir on a larger scale; but that time never came.

This fuller "Memoir" is now attempted by an inferior hand; yet I am pleased to have this opportunity of carrying out, to the best of my abilities, an intention of one for whom I feel such warm regard and respect, and with whom I had held friendly and intimate relations during the quarter of a century previous to his death, in 1890. I know that he would have regarded my work with kindness and its shortcomings with considerate allowances, while he would have warmly sympathised with my object of preserving or reviving the memory of him to whom he referred as "among the *Dii majorum gentium* of scientific history."

I have received much kind assistance in my work, without which I could not have undertaken it. Mrs. Rennell Rodd—the widow of Major Rennell's only grandson—placed at my disposal her husband's manuscript volume of family history, containing numerous memoranda of great interest, several letters from India, an important letter from Sir Edward Sabine, the series of letters from Major Rennell to his grandson, and the copy of Baron Walckenaer's *éloge*, originally sent to Lady Rodd. Mrs. Rennell Rodd also lent me the Thackeray family history by Mrs. Bayne, privately printed, and containing the memoirs of Mrs. Rennell's family, with notices of Major Rennell and his children. To Mr.

F. Edmund Langley, of Chudleigh, I am indebted
for ninety-four letters from Rennell, chiefly to his
guardian, the Rev. Gilbert Burrington, of Chudleigh,
from 1758, when he was a midshipman, aged sixteen,
to the year 1785. I have to thank Earl Spencer for
allowing me to peruse several letters to his grandfather, preserved among the muniments at Spencer
House, and Mr. Morris Beaufort for the use of fourteen
letters to his father, Sir Francis Beaufort. Letters to
Admiral and Mrs. Smyth were kindly lent me by
their daughter-in-law, Lady Warrington Smyth. I
am also obliged to Sir William Flower for drawing
my attention to a letter from Major Rennell to Dr.
John Hunter, and for lending me his copy. Mr.
William Foster, of the India Office, Secretary to the
Hakluyt Society, has been so obliging as to search
and make extracts from the records, wherever Major
Rennell's name occurs, from 1778 to 1817 ; and I have
also examined the manuscript maps and field books
at the India Office, and all Rennell's charts and plans
engraved in the collections of Dalrymple both at the
India Office and in the map-room of the Geographical
Society.

My other authorities are the works of Rennell
himself and of his critics, and the histories and
memoirs of the time. Parts of the opening pages of
the last chapter, giving an account of the origin of
the Raleigh Club, are taken from my "History of the
First Fifty Years of the Geographical Society," written
in 1880.

CONTENTS.

MAJOR JAMES RENNELL

AND THE

RISE OF MODERN ENGLISH GEOGRAPHY.

———•:•———

CHAPTER I.

HOME AT CHUDLEIGH.—SERVICE IN THE NAVY.

THE answer to the question "Who was the first and greatest of English geographers?" can be made with confidence. James Rennell may not have been the father of English geography, but he was undoubtedly the first great English geographer. Much laborious work had to be done before geography became a science in this country. Materials had to be collected, instruments and projections had to be invented and improved, correct methods of criticism and accurate habits of thought gradually had to be established, before the work of the scientific geographer could commence.

The fathers of English geography were Richard Eden and Richard Hakluyt, who in 1555 and 1589 published the first collections of voyages and travels. They stimulated the love of adventure, and encouraged the spirit of enterprise which has ever

since continued to supply geographical food, and
all the incalculable advantages to the country which
such nourishment ever produces. Side by side with
the supply of knowledge must advance the means of
obtaining it. Scientific measurements are essential to
accurate geographical information. While Hakluyt
was collecting his records, Davis was inventing his
quadrant, Molyneux was constructing his globes; and
somewhat later, Wright was utilising the projection
of Mercator. A little later still Purchas made his
Pilgrimes tell their marvellous tales of adventure by
sea and land; while Napier invented logarithms,
Henry Briggs and Edmund Gunter brought them into
practical use, and Hexham furnished English readers
with the atlas of Hondius.

Thus the accumulation of geographical knowledge
advanced hand in hand with the science of accurate
measurement and delineation; and in the twenty
years previous to the birth of Rennell, Hadley had
invented his quadrant, Harrison had constructed his
first chronometer, and the earlier circumnavigations of
the eighteenth century had been projected. The
publication of the Nautical Almanack, by Nevill
Maskelyne, was not commenced until twenty-five
years after Rennell was born. It was necessary that
knowledge should be accumulated during a lengthened
period, that great advances should be made in the
perfecting of scientific appliances, and that the critical
faculty should have been cultivated by deductive
reasoning, before there could be modern successors to
Eratosthenes and Hipparchus, to Strabo and Ptolemy.

A geographer is so many-sided that it is not easy
to give a comprehensive definition of the term. He
should have been trained by years of land or sea

surveying, or both, and by experience in the field in delineating the surface of a country. He should have a profound knowledge of all the work of exploration and discovery previous to his own time. He should have the critical faculty highly developed, and the power of comparing and combining the work of others, of judging the respective value of their labours, and of eliminating errors. He must possess the topographical instinct; for, like a poet, a geographer is born—he is not made. He must be trained and prepared for his work, but he must be born with the instinct, without which training and preparation cannot make a finished geographer. He must have the faculty of discussing earlier work and of bringing out all that is instructive and useful in the study of historical geography; and he must have a competent knowledge of history and of sciences which border upon and overlap his own, so far as they are understood in his day. Such men do not arise until the time is ripe for them. James Rennell possessed all these qualifications. He also had the advantage of succeeding to the labours of earlier foreign workers in the same field. De L'Isle and D'Anville had gone before him. They were the first French geographers, and Rennell was aptly called " the English D'Anville."

The birthplace and youthful surroundings of our first English geographer will, therefore, have a special interest for the student of his life-story; for we thus become acquainted with the images which filled his brain at the time when his geographical instincts first began to develop themselves.

In driving along the high road from Exeter to Plymouth for ten miles, the traveller comes to the ancient market-town of Chudleigh, on the banks of

the little river Teign: "lying under the Haldon Hills
to the west," as old Westcote describes it.   About a
mile from Chudleigh there were two freehold proper-
ties, called Waddon and Upcot, owned by Captain
John Rennell, of the Artillery, who was married in
1738 to Ann Clarke, of Chudleigh.   The Rennells had
been inhabitants of Chudleigh for many generations.
It is probable that they descend from the Reynells of
Plantagenet times.   Waddon had been in the family
since the days of Queen Elizabeth, but the home of
the captain of the artillery and his wife was at a house
built by his grandfather at Upcot.   John and Anne
Rennell had two children—Sarah, born in 1740, and
James, born at Upcot on December 3rd, 1742, and
baptized by Mr. Bayley, the vicar of Chudleigh, on
the 21st of the same month.

The earliest thing that James Rennell could remember was parting with his father at Woolwich
when he embarked for the wars.   Soon afterwards
the news of the captain's death arrived at Chudleigh.
He had fallen in some action, the name of which is
not recorded,* about two years after the battle of
Fontenoy.   In July, 1747, Mrs. Rennell was left a
widow with two children, and her affairs were so
embarrassed that the property had to be sold.   At first
she found a home with a distant cousin of her husband,
Dr. Thomas Rennell, the rector of Drewsteignton,
near Exeter.   Dr. Rennell, who was father to the Dean
of Winchester, used to say in after years that he
taught little James to read ; but the boy had no
very pleasant memories connected with Drewsteignton
and his Barnack cousins, as he called them.   Before

---

* Baron Walckenaer says it was at the Battle of Lawfeldt.

very long Dr. Rennell succeeded to the living of
Barnack, in Northamptonshire, and the widow had
to seek another home. She married a Chudleigh
neighbour, named Elliott, a widower with a family by
a former wife, and very limited means. They lived
at Exeter, and the daughter, Sarah, was received by
her step-father; but he could not afford to keep the
boy or to give him much help. When Mr. Elliott
died, twelve years afterwards, his step-son Rennell
wrote of him :—" I am persuaded Mr. Elliott's heart
was very good. He only wanted the means to
serve me."

James Rennell was but ten years old, and he was
left in a forlorn position. It is believed that he was
at Pynsent's Free Grammar School at Chudleigh for
a short time. Yet this was the commencement of
a very happy period in his young life. Mr. Bayley
had died, and had been succeeded as Vicar of
Chudleigh, in 1752, by the Rev. Gilbert Burrington.

The new vicar's family consisted of his wife, who
was a Miss Savery, his sister Miss Burrington, an
old relation named Mrs. Sampson, and a baby, which
was quickly followed by two others. These warm-
hearted people received the fatherless boy as their
own child. He cordially returned their affection,
and with them James Rennell's life became a very
happy and pleasant one.

There were many associations connected with
Chudleigh, as the home of his ancestors, which were
likely to fix themselves on the boy's mind, and all
the surroundings would be calculated to stimulate
his youthful geographical tendencies. The old
church, built five centuries before his time, would
form the central point of his survey. Inside were

the effigies of grim Sir Piers Courtenay and his wife,
ever kneeling side by side, and the Prophets and
Apostles quaintly painted on the panels of the
ancient screen. Without, the old tower rose above
the vicarage garden, and over the street down which
the Plymouth coach drove every day, with the guard
sounding his horn, and waking up the little town
from its slumbers. Farther afield there was many a
scene of enchantment for a young boy. The surrounding country is intersected by a great number
of those deep and solitary lanes, with their banks of
ferns and wild flowers, which are so characteristic of
Devonshire. A mile north of the town are Waddon
and Kerswell Rocks, and on the other side are all the
sylvan beauties of Ugbrooke Park, with its "castle
dyke," its noble clumps and avenues, and the stately
grove of beech-trees, the favourite walk of the poet
Dryden. Most beautiful of all was the Chudleigh
Rock, rising in a perpendicular cliff, and seen through
breaks in a wild and tangled wood, where a noisy
stream rushes and eddies among moss-grown stones,
and in one place falls in a creamy cascade. The
mouth of a dark cavern is seen on the face of the
cliff, and from the summit of Chudleigh Rock there
is a glorious view of the ridge of Dartmoor, broken
by the granite masses of Hey-Tor and Rippon-Tor.

Amidst these scenes James Rennell passed the
happiest days of his boyhood, his spirits brightened
and his heart softened by the constant affection and
care for him shown by the family at the vicarage.
Even in these very early days he seems to have
looked at the scenery around him with the eye of a
young geographer; and there is a tradition handed
down by the Bishop of Barbadoes, who married one

of the Barnack Rennells, that young James constructed a plan of the country round Chudleigh at the age of twelve.

The Plymouth coach would keep up some sort of interest in naval matters at Chudleigh; the geographical instinct would be another originating cause; and thus the thoughts and aspirations of James Rennell were turned seawards when he was approaching the age of fourteen. The vicar had an opportunity to help his young friend to attain his desire. His brother-in-law, Mr. Savery, of an old Totnes family, was a retired barrister, living at a place called Slade, near Ivy-bridge. He was a friend of Captain Hyde Parker, who was commissioning the *Brilliant* frigate, and thus it was that Rennell obtained his first naval appointment, with the rating of captain's servant, in January, 1756.

England was on the eve of the seven years' war with France, which was declared on the 18th of May, 1756. Those were the days when Lord Anson, the circumnavigator, was First Lord of the Admiralty, and when Hawke and Boscawen commanded our fleets. Life in the navy was much rougher than it is now. There was seldom anything but ships' provisions in the midshipman's berth, and these provisions were not so appetising as they became after the Mutiny of the Nore. The weekly accounts of those days give bread, beef, pork, pease, oatmeal, butter, cheese, beer, with salt fish on banyan days. The reports of surveys on provisions, especially on cheese and beer, were revolting. Pea-soup and lobscouse were tolerable, but dog's-body and burgoo were merely filling, without relish. Clothing was not fixed by any rules, except for commissioned officers. Before Lord Anson's time,

indeed, the lieutenants purchased the soldiers' old coats
at Gibraltar or Port Mahon, trimmed them with black,
and wore them as uniform. In 1747 the Admiralty
issued orders that a uniform should be worn, consisting
of a blue coat with white collar, cuffs, and facings ; yet
as late as 1759 young Rennell saw a master of a king's
ship wearing a red coat with black facings, and thinking himself very smart. It may have been an old
coat of one of the lieutenants, who then wore blue
uniforms ; the ranks being distinguished by gold lace,
and the midshipmen having those picturesque patches
with button and twist, which still survive.

But although the life in a midshipman's berth was
rough, it was not always unpleasant. In spite of
noise and interruption, there was reading and study,
as well as agreeable intercourse. Then, as now, the
truest and most enduring friendships were formed
amongst midshipmen, as we shall see in the course of
the present biography; and the midshipman's berth
has been the nursery not only of the naval preservers
of our country, but also of renowned generals and
learned chancellors. The poet Falconer was the contemporary of young Rennell, and has graphically
described the midshipman's berth.*

> " In canvass'd berth, profoundly deep in thought,
>   His busy mind with sines and tangents fraught,
>   A Mid reclines : in calculation lost
>   His efforts still by some intruder crosst.
>   Now to the longitude's vast height he soars,
>   And now formation of *lobscouse* explores.
>   Now o'er a field of logarithms bends,
>   And now to make a pudding he pretends ;

* Written a few years before 1762 : probably in 1758.

At once the sage, the hero, and the cook,
He wields the sword, the saucepan, and the book.
Opposed to him a sprightly messmate lolls,
Declaims with Garrick* or with Shuter† drolls ;
Sometimes his breast at Cato's virtue warms,
And then his task the gay Lothario charms.
Cleone's grief his tragic feelings wake,
With Richard's pangs the Orlopian‡ caverns shake.
No more—the mess for other joys repine
When pea soup entering shows 'tis time to dine.
But think not meanly of this humble seat
Whence spring the guardians of the British Fleet.
Revere the sacred spot, however low,
Which formed to martial acts a Hawke, a Howe."

Here we have the picture of a midshipman trying
to work out his sights in the intervals of amateur
cooking and in the midst of much uproar; while
another, sitting opposite to him, declaims snatches out
of old plays; both occupations instantly ceasing on
the arrival of the pea soup.

Before entering upon these exciting scenes, young
Rennell had to take leave of his friends at the vicarage. The Rev. Gilbert Burrington continued to be
his adviser, and always acted as his agent. Rennell
constantly corresponded with him from the time of
his entering the navy until the old vicar's death, thirty
years afterwards. He thought and spoke of the vicar
and his family with the tenderest affection, and when
they were mentioned, even by a stranger, his eyes
filled with tears of gratitude.§ Many were the

* David Garrick was born in 1716. He was then in the height of
his fame, having taken Drury Lane in 1747, and retired from the
stage in 1776. He died in 1779.
† A celebrated comic actor in those days. He died in 1776.
‡ The cockpit, which was on the *Orlop* deck.
§ Letter from Mr. Topham to Mr. Burrington.

B

messages he sent to Mrs. and Miss Burrington and to
old Mrs. Sampson ; and the midshipman's letters in-
variably ended with "love to the dear children."
Tom, Robert, and Gilbert Burrington were their
names, and Tom was old enough to write to his midshipman friend before he finally left England. When
Tom reached the age of fourteen, young Rennell was
already receiving a good salary in India. He sent
the boy fifty guineas through his father, with these
words :—" Thy father was my friend ; let me be a friend
to thee for ever." No lapse of time could efface from
the mind of young Rennell the deep debt of gratitude
he owed to the Burringtons, to whom was due all the
brightness and happiness of his boyhood at Chudleigh.
We can picture them to ourselves crowding round the
Plymouth coach outside the vicarage gate, and waving
their farewells, while the young midshipman climbed
up and was driven away.

James Rennell joined the *Brilliant* frigate (Captain Hyde Parker), in January, 1756, and during the
two following years he was cruising on the coast of
Spain or in the Channel. He received great kindness
from the captain, and was happy with his messmates.
One of them, a midshipman named Topham, was his
great friend. Writing to Mr. Burrington years afterwards, Rennell said of himself and Topham :—" In
our former humble station we always endeavoured
to promote each other's happiness, when we lived
together as messmates for nearly three years." With
such friendships the duties on board a frigate and
the life in a midshipman's berth were made very
agreeable.

More active service for the *Brilliant* commenced
when she was ordered to join the squadron under the

command of Commodore Howe, in 1758. While
Lord Anson watched the French fleet in Brest, troops,
under the Duke of Marlborough and Lord George
Sackville, were embarked on board Howe's ships in
June, 1758, and landed in Cancalle Bay, five miles
from St. Malo, on the 6th. After burning a fifty-gun
frigate and seventy privateers in the river Rance, the
troops were re-embarked on the 9th without accident.
The Duke and Lord George then went to the war in
Germany, and an old officer named General Bligh
took command. On August 1st Lord Howe* was
at St. Helens, ready for sea, with his broad pendant
on board the *Essex*. Sir Edward Hawke had relieved
Lord Anson in the blockade of Brest. The earliest
letter from young Rennell that has been preserved is
dated on board the *Brilliant* at St. Helens on July
2nd, 1758. It gives an account of the landing at
Cancalle, and the midshipman expresses himself
greatly obliged for advice to keep good company,
which he hopes he will follow.

On August 6th Howe's squadron was off Cher-
bourg, and the troops were landed under Colonel
Dury, of the Guards, who entered the town without
opposition, blew up the forts, spiked a vast number
of guns, and destroyed the piers and basin. The
harbour was rendered useless, and the troops were
successfully embarked without any loss. The
squadron then encountered a gale of wind, and put
into Portland on August 17th to refit, but early in
September Lord Howe proceeded on a further enterprise, which was directed against St. Malo. Prince

---

* He succeeded to the title by the death of his brother on
July 3rd, 1758.

B 2

Edward, afterwards Duke of York,* then aged nineteen, was serving on board the *Essex*.

This time the troops, under General Bligh, were landed in the Bay of St. Lunaire, in Brittany, on the left bank of the Rance. After having reconnoitred St. Malo, the general abandoned the idea of attacking so strong a place, and decided to re-embark. But in the meanwhile, the Bay of St. Lunaire had been found to be too exposed, and Lord Howe selected the Bay of St. Cast, further to the westward, where there was a good sandy beach and better anchorage, for the place of embarkation. General Bligh was, therefore, requested to march by land to the Bay of St. Cast, where Lord Howe made the necessary preparations. Five sloops and bomb-ketches were anchored, in line, as near the beach as possible, and six frigates were placed farther out: namely, the *Pallas*, with the flag of Lord Howe, shifted from the *Essex*, the *Brilliant*, *Montague*, *Portland*, *Jason*, and *Salamander*. General Bligh commenced his march to St. Cast, his troops being harassed by sharp-shooters as they passed the woods and villages. At Martignon, the general received news that the French, under the Duc d'Aiguillon, were in great force between that village and St. Cast: upwards of 10,000 men.

On Monday, the 11th of September, 1758, at early dawn, everyone on board the frigates was anxiously on the look-out for troops. Young Rennell, on board the *Brilliant*, was ready to take bearings and make a plan of the scene of operations; for his

* Brother of George III. He was created Duke of York and Albany in April, 1760, and became a rear-admiral at twenty-one. He died unmarried in Italy, aged twenty-eight, on September 17th, 1767, and was buried in Westminster Abbey.

geographical instincts appear to have turned his attention to surveying and to the construction of plans and charts from the moment of his entering the navy.

At about eight o'clock in the morning the little English force first appeared on the crest of the hill, near two windmills, and the bearings were carefully taken by Rennell; he watched the red line marching straight down the slope to embark. Boats were immediately sent on shore, and several captains of the frigates landed to superintend the embarkation. Lord Howe, with Prince Edward, was also away in his galley. Very soon after the appearance of the British troops the French came in view a little to their right, and planted a battery of six field-pieces, with a view to harassing the march of the British to the beach. But they were soon driven from this position by the fire from the frigates and the bomb-ketches, and they took refuge in the village of St. Cast and among trees half-way down the slope, but farther to the right, where they could not be seen.

The embarkation then proceeded without molestation, until only a rear-guard of 1,200 men, including a detachment of the Grenadiers under Colonel Davy, remained on the beach. Then it was that the French, in overwhelming numbers, issued forth from the village, which was at a distance of about half-a-mile, and attacked the little English rear-guard with great fury. The Guards retreated behind a dyke, which Rennell describes as having been thrown up by the peasantry to prevent the sea, at spring-tides, from overflowing the low grounds. The French masses of troops were thus exposed to the fire from the sloops and bomb-ketches, and might have been forced to retreat. But Colonel Davy could not endure the sight of the poor men on the beach and in the boats exposed to be slaughtered by heavy odds. Unfortunately, he sallied forth from behind the dyke, and led the Guards against the enemy. He fell gloriously, but his men were overwhelmed by superior numbers, more than five to one; and there was a sickening scene of slaughter, the gallant fellows being driven into the sea and ruthlessly shot down; while the vessels were obliged to cease firing because friends and foes were inextricably mixed. The French brought down their field-pieces, and used them with deadly effect. A small remnant received quarter, and were taken prisoners, including Lord Frederick Cavendish, and as many as five of the captains of frigates who were superintending the embarkation. Lord Howe remained close to the beach in his boat until the very last, giving directions and encouraging the boats' crews by his example. There were 700 men missing out of the 1,200 on the beach when the attack began, of whom 500 were prisoners.

Young Rennell was busily engaged during the action in taking notes and bearings for the purpose of preparing a plan of St. Cast Bay, showing the positions and movements of troops and ships, and the surroundings of this disastrous action. " Plan of St. Cas Bay; J. Rennel, fect., 1758. To the Right Honourable Lord Howe this plan is dedicated by his obedient servant, J. Rennel." It is very neatly drawn, and a copy, which has been preserved, was sent to his friend Mr. Burrington. It is the earliest specimen of Rennell's work as a surveyor, but it was drawn at a time of great excitement and sorrow. " It was a very shocking sight," he wrote, " to me and the rest who saw all that was done, to see our poor soldiers running some one way and some another, some into the sea to escape the enemy, who shot them down without mercy. None of the frigates were far enough in to fire on the enemy, save the *Saltash* and *Swallow* sloop."

One result of the action was that Captain Parker was transferred to the *Montague* to replace Captain Rowley, who was taken prisoner at St. Cast, and the *Brilliant* was given to Captain Sterling, of the *Essex*. Returning to Devonport, young Rennell was very busy at the work of refitting the *Brilliant* during March, 1759, and getting her ready for a cruise to the westward, in company with the *Deptford*. This cruise lasted during the month of April, and it was a very successful one. When off Scilly Islands, the *Brilliant* came in sight of a large privateer, overhauled, and captured her. She proved to be the *Marquis du Baraille*, of Dunkirk, mounted with fourteen six-pounders and with a crew of 120 men. A few days afterwards the *Basquaise* was also taken, a Bayonne privateer, mounting twenty-four six-and nine-pounders, with a crew of 220 men. This second capture obliged the *Brilliant* to return to Devonport, for she had more prisoners on board than captors, and was running short of provisions. During the summer there was another cruise, but in the autumn the young midshipman got leave to see his friends. He was no doubt at Chudleigh during part of the time, but we only hear of him at Slade, near Ivybridge, staying with his friends Mr. and Mrs. Savery, in November, 1759.

Captain Hyde Parker had been appointed to the *Norfolk*, seventy-four-gun ship, to join the fleet in the East Indies, and he had consented to take young

Rennell with him if he could get round to Portsmouth
in time. It was with the hope of more fully securing
the approbation of Captain Parker that the lad volun-
teered for the East Indies, for interest was the only
road to advancement in those days. In a letter to
Mr. Burrington, he says :—" I flatter myself that my
going with Mr. Parker so long and hazardous a voyage
will get me the more into his interest; nor is it with
any other view than that that I shall undertake it.
It is not in the least unlikely that he will be made an
admiral before he returns, and should a vacancy occur,
he can promote whom he pleases."

The young midshipman's great anxiety was to get
a passage round to Portsmouth to join the *Norfolk*,
and Captain Lucas had promised to take him in the
Torrington. In December, 1759, he received his
prize-money for *La Basquaise*, and spent a portion in
the purchase of what he considered necessary books.
" I imagine," he wrote, " that we shall lay siege to
Pondichery or some other French settlement, and
have bought some books which are very needful for
me. They are Muller's works, which will give me a
perfect idea of attack and defence, and the method of
choosing ground, building forts, and taking plans of
places. These are two large volumes, which, with
some other useful books for the sea, amount to one
guinea." Thus provided, he awaited the sailing of
the *Torrington*, but she got under weigh on the night
of January 2nd, 1760, without making any signal. In
frantic haste, he boarded the *Supply* transport, one of
the *Torrington's* convoy, with his traps, and kept
company with the frigate during the next two days.
But in the hurry there was some informality in his
leaving the *Brilliant*, and he was entered " run " on

that ships books, which caused him great annoyance, and gave Mr. Burrington much trouble in recovering his pay and the rest of his prize-money.

On Friday, the 4th of January, during a fresh gale, with thick hazy weather, the *Supply* ran on shore on a ledge of rocks near the north-west end of the Isle of Wight. It was very dark, the ship was beating on the rocks for about two hours, the wind increased to a whole gale, and there was a high sea. Eventually the force of the wind hove her off at about seven in the morning, and she was anchored off Bembridge. Besides Rennell, there were four midshipmen on board and six women, who had been left behind by the *Torrington*, and whose screams were dreadful while the ship was bumping on the rocks. The *Supply* got safe into Portsmouth Harbour on Saturday, the 5th of January, but to Rennell's dismay, he found that Captain Parker had already sailed. The only thing to do was to follow him in some other ship. Almost distracted, the boy hurried to the port admiral, and entreated that he might be allowed a passage in some other ship destined for the East Indies. But there were admirals, both in those days and since, who treated midshipmen as if they were beneath their notice. Admiral Holborne was one such. "I am sorry to say," wrote Rennell, "that his behaviour to me was very unbecoming a gentleman." Rennell next applied to Captain Haldane, who was in command of the *America*, a frigate at Spithead, about to sail for the East Indies. This officer assured the young applicant that he would be extremely welcome to go on his quarter-deck, and that he would get the necessary order from the admiral. In the beginning of February, 1760, young Rennell was

installed as a midshipman on board the *America*,
having furnished himself with drawing compasses,
a navigation book, and a Hadley's quadrant. But
he felt the loss of the companionship of messmates
who had become real friends on board the
Brilliant. Out of the twelve midshipmen on board
the *America*, he could only pick out three who
were conversible and likely to become friends; but
the surgeon's mate was not only a Devonshire
man, they had mutual acquaintances about whom
they could talk: which is always a great comfort on
board ship.

It was in a letter from the *America* that young
Rennell promised his guardian always to spell his
name with two ll's in future.

All through February there was very bad weather
at Spithead, and the *America* remained at anchor.
On the 28th the melancholy news of the loss of the
Ramillies arrived. Admiral Boscawen had sailed
from Plymouth on the 7th to join Sir Edward Hawke
in Quiberon Bay with a small squadron. A violent
gale dispersed the ships, and the *Ramillies*, in trying
to make Plymouth in thick and hazy weather, passed
the port and got embayed near Bolt Head, which was
mistaken for the Rame. Unable to weather the point,
the captain ordered the masts to be cut away, and
anchored. The cables parted, and the ship was driven
on the rocks, and dashed to pieces. All hands were
drowned but one midshipman and twenty-five men.
The sad tidings reached Portsmouth when the *America*
had just suffered serious damage from the effects of
the same gale of wind. She was quickly refitted, and
was so filled up with stores and provisions, the lower
deck being full and the ports caulked in, that she

looked like a loaded merchant ship. On the 6th of March she sailed for Madras.

Young Rennell already had ideas of entering the East India Company's service, if interest in the navy failed him. While the *America* was at Spithead one of the midshipmen went up to London to pass for a lieutenant, but not having any interest, he was turned back for some trifling reason. Next day he went to the India House, gave the directors an account of what had happened to him, and applied for permission to pass their examination for a mate : which he did. He was immediately sent on board one of their outward-bound ships as second mate. The fate of his messmate set Rennell thinking, and his conclusion was that the post of second mate of an Indiaman was far preferable to that of any naval lieutenant. Still, he was full of hope from Captain Parker's patronage, and when she sailed the *America* promised to be a very pleasant ship. " I think," he wrote, " that I never lived happier since I have been in the service than in this ship. The lieutenants are all very young, but behave extremely well. If I have my health, I have not the least doubt of preferment this voyage ; indeed, I have more hope of it than ever."

The voyage out occupied six months, and Rennell kept in excellent health, though many of the crew were afflicted with scurvy. They touched at Madeira, and reached Madagascar in July, when they were eighteen weeks from England. Here the *America* remained a fortnight, and the scurvy-stricken people soon recovered, with abundance of fresh meat, fruit, and vegetables. Young Rennell was puzzled at his own complete freedom from scurvy. He had never eaten more plentifully of salt provisions since he had been

at sea, and he made use of very little acid ; while the
ship must have been in a most unwholesome state.
She was so deep in the water that the ports could
never be opened, and was very close on the lower
deck. The only way in which he could account for
his exemption was that he kept as much on deck and
aloft, and took as much exercise as possible.

The only drawback to the pleasure of the voyage
was the outrageous conduct of Captain Haldane. He
was exceedingly civil and agreeable at Portsmouth
and Spithead, but as soon as they entered the tropics
he came out in his true colours, and developed a most
ungovernable temper. He used a stick to some of the
midshipmen, and actually thrashed young Rennell
with his fists before all hands, treating a messmate of
like age, a nephew of the Speaker Onslow, in the
same way. Rennell was unable to forgive such an
insult. " I can forgive most injuries," he wrote, " but
not of this nature ; nor shall my utmost endeavours
be wanting to resent it, if ever I return to England in
a capacity which will enable me to do so." A week
afterwards the captain sent for the two midshipmen
to make a sort of apology, telling them that he was
cursed with a bad temper ; but this did not abate their
resentment. Poor Captain Haldane died at Bombay
in April, 1761 ; and Rennell only wondered how a man
of so turbulent a temper had lived so long.

As soon as the *America* arrived at Madras,
Captain Hyde Parker behaved to his old shipmate
like a sincere friend. Admiral Stevens had hoisted
his flag on board the *Norfolk*, and Captain Parker
had been transferred to the *Grafton*, a third-rate of
sixty-eight guns. Rennell was at once appointed to
the *Grafton*, and he wrote home in very good spirits.

" I can now say that I enjoy the happiness of a good ship, captain, and officers, and very agreeable company. Perfect harmony subsists among the officers, and excellent discipline among the men. I do my duty with pleasure, and enjoy my leisure moments in peace."

James Rennell was now a midshipman of some standing, having seen a good deal of service. He was eighteen years of age, and within a short time of passing for a lieutenant. He was a well-conducted young officer, zealous and intelligent, considerate for others, and with a great capacity for making friends. His love for geography had been developed by study, and by his opportunities for surveying and drawing charts. This was a preparation for more important work. His admirable training in the navy was destined to lead to great results, and to his employment in India during nearly twenty years of arduous and distinguished service.

CHAPTER II.

WHEN James Rennell arrived in India, in September, 1760, the English had already entered upon possession of Bengal, Bahar, and Orissa. The battle of Plassy had been fought three years before, and in February, 1760, Lord Clive had returned to England, at the termination of his first government of Bengal. Mr. Vansittart had succeeded him at Calcutta; and in the same year Sir Hector Munro won the battle of Buxar over Sujah Daulah and Mir Kosim. In Madras the French ascendency acquired by Dupleix and Bussy was rapidly waning, under Count Lally. In 1758 Lally had sent 2,000 men, under D'Estaing, against Fort St. David. The place surrendered on June 2nd, and was reduced to ruins. In December Lally himself invested Madras, but the siege was raised in February, 1759; and this was the last gleam of success enjoyed by the French in India. Colonel Coote arrived at Madras, and in January, 1760, he defeated Lally at the battle of Wandiwash, and captured Arcot, Chingleput, and several places round Pondichery, including Carical. In August 400 marines were landed at Cuddalur, and the capital and centre of French dominion, their settlement of Pondichery, was soon afterwards hemmed in on every side.

Admiral Watson was with Clive when Calcutta was recovered in 1757. He died in August, 1757, and in February, 1758, Admiral Pocock succeeded to the command. He was opposed to a French fleet, under Count d'Achè. There was an indecisive action off Carical in August, 1758, when Commodore Stevens, the second in command of the English fleet, was wounded in the shoulder. But the French showed much prudence throughout these operations on the coast of India, avoiding an action, and D'Achè passed most of his time at Mauritius. On Sir George Pocock's retirement, Stevens, who had become rear-admiral in December, 1758, hoisted his flag on board the *Grafton*, shifting it to the *Norfolk* in 1760. His second in command was Captain Samuel Cornish, who began life as an apprentice on board a collier. He entered himself as a seaman in the navy, and was soon made a boatswain. Young Cornish owed his future advancement to his own intrinsic merit, obtaining the rank of a commissioned officer, and in 1742 being captain of the *Nassau*. He came out to the East Indies in May, 1759, with his broad pennant on board the seventy-four-gun ship *Lenox*, as second in command, first under Sir George Pocock, and then under Admiral Stevens.

Captain Parker, with whom Rennell had already served for three years on board the *Brilliant*, was a son of the Rev. Hyde Parker, rector of Tredington, in Worcestershire, and grandson of Sir Henry Parker, the first baronet, by Margaret, daughter of Dr. Hyde, Bishop of Salisbury, a relation of the Chancellor. Hyde Parker was a master's mate in the *Centurion* with Commodore Anson in the voyage of circumnavigation, and became a captain in 1748. He was a very distinguished officer, and the *Grafton* was in excellent order.*

Young Rennell's first letter home from India was dated on board the *Grafton*, in Fort St. David Roads, on September 30th, 1760.   Everything was new and strange to him.   He described how the entire dress of the Europeans, except hats and shoes, was of linen, and how the shoes were made of tanned skins no thicker than coarse paper.   The desolate condition of Fort St. David aroused his anger.   " The French pillaged it and destroyed both houses and fortifications, so that it is now quite desolate and uninhabited.   This vile proceeding of the French, who had the ambition to think of driving us out of the Indies, is justly resented by every person here, and the delinquents are now cooped up within the walls of Pondichery.   Their fleet has not offered since their action about twelve months ago.   They were then equal to ours, but we have since been reinforced by four sail-of-the-line and two frigates.   Our whole force by sea consists of two ships of seventy-four, one of sixty-eight, three of sixty-four, eight of sixty, two of fifty guns, and six frigates."   He also speaks very highly of Admiral Pocock's arrangements for provisioning

* Captain Hyde Parker afterwards served under Admiral Cornish at the taking of Manilla, and captured a Spanish galleon valued at £500,000.   In 1778 he became a rear-admiral, and served on the North American station.   He was on board the *Victory*, in command of the North Sea fleet, in 1780, and had an encounter with the Dutch fleet on the Dogger Bank.   In 1781 he succeeded to the baronetcy, on the death of his clergyman brother without children.   In April, 1782, he was appointed commander-in-chief in the East Indies, and went out in the *Cato*, but she was never heard of again after leaving Rio in December.   Sir Hyde Parker's son, also Sir Hyde Parker, was Nelson's senior officer at the battle of Copenhagen.

the fleet. "By his industry and care he so far improved the manner of living that the fleet is well supplied with almost every article. Fresh beef and soft tack are issued to the crews every day, and fruit and vegetables are very plentiful."

The capture of Pondichery was the final blow to all chance of French ascendency. By October, 1760, Colonel Coote had completed the blockade by land, and Count Lally had cruelly driven all the inhabitants out of the town. The siege then commenced, and Admiral Stevens co-operated by sea, making Cuddalur his rendezvous.

On the 6th of October an expedition was organised for cutting out a large French frigate and an Indiaman at anchor under the guns of Pondichery. Two unsuccessful attempts had already been made. This time two armed boats were told off from every ship in the fleet, which was then at anchor at Cuddalur, about eighteen miles to the south of Pondichery. The boats started at sunset, and rowed all night. Young Rennell was selected as a volunteer in the division commanded by Lieutenant Ouvry. Both the ships were at anchor within half a musket-shot of the town, and in such a position as to receive protection from the bastions. At about two in the morning the English sailors boarded and cut the cables, and at the same moment the garrison was alarmed, and soon a furious cannonade began from every gun on the walls of Pondichery that could be brought to bear on the two ships. The ship which Rennell boarded had no sails bent, and was exposed to the fire of the enemy for a full hour, until she could be got ready for sea. Yet the whole British loss was only eight killed and thirty wounded. The frigate was called the *Baleine*,

C

and was soon afterwards put in commission. The Indiaman, named the *Hermione*, was in ballast.

The army remained in camp during the rainy season in order to push the siege, and Admiral Stevens left five ships of the line to continue the blockade by sea. With the rest of the fleet, he proceeded to Trincomali, in Ceylon, arriving on November 1st. Rennell pronounced Trincomali to be one of the best harbours in the world, and he at once conceived a desire to make a survey of it, in which laudable project he was kindly encouraged and assisted by Captain Parker. He had made some plans of harbours and anchorages on the passage out, which he presented to his captain; and he hoped that his plan of Trincomali would go some way towards gaining the favour both of Captain Parker and Commodore Cornish. He had found means of mastering the theory and practice of marine surveying, and, in accordance with Mr. Burrington's advice, he announced his intention of practising it whenever he had an opportunity.

The admiral, impatient to return to Pondichery, left Trincomali in the middle of December, before the bad weather was over. The fleet had only returned a few days when a furious hurricane burst upon it. Luckily, the *Grafton* was lying farthest from the shore. The ships were soon parted from their anchors by the breaking of the hempen cables, and driven before wind and sea half full of water, those on board expecting every moment to be their last. Their masts were either blown or cut away, and nothing was thought of but immediate death. The *Duc d'Aquitaine* sank half a mile from the shore, only one man being saved. The *Sunderland* and an

Indiaman sank a little farther off. The *Newcastle, Queenborough,* and *Protector,* fire-ships, were driven on shore and bilged. The *America, Medway, Panther, Falmouth,* and *Liverpool* were all dismasted. The *Norfolk* and *Grafton* were the only two ships, out of thirteen, that saved their masts. This disaster reduced the fleet to ten ships, two of which were scarcely seaworthy. Jury masts were rapidly fitted, in case the enemy's fleet, hearing of the catastrophe, should venture to put in an appearance.

Still the blockade of Pondichery was continued by sea and land, and on January 17th, 1761, the place surrendered by capitulation, after a siege of five months. Rennell said that this was due to want of provisions, and not to any fear of the place being taken by storm, for the besieging batteries had not done the French the least damage since they opened fire.

The fleet then proceeded to Bombay, arriving in February, 1761. Rennell had learnt to swim at Trincomali, and when he was practising in Bombay Harbour he was so severely stung by some marine animals that fever supervened, and he was very ill for several months. He lived on shore while the ship was in dock, and wrote home a very graphic and interesting description of Bombay to Mr. Burrington. Poor Admiral Stevens had grown so very fat during this service in the tropics that his life had become a burden to him, and at last he died of apoplexy on May 2nd, 1761, much lamented, for he was an excellent officer. He was succeeded in command of the East Indian station by Admiral Samuel Cornish, a thorough seaman, a very capable commander-in-chief,

c 2

and an intimate friend of Captain Hyde Parker.
Young Rennell's six years' service would be completed
early in January, 1762, when he would pass for a
lieutenant, and he thought that his chances of promotion were still good so long as the war continued.

Admiral Cornish matured a plan for capturing
the islands of Bourbon and Mauritius. He sent his
scheme to the Admiralty, and during the last half of
1761 he was busily engaged at Bombay in making
preparations. Flat-bottomed boats for landing troops
were finished, transports were taken up for conveying
part of the garrison of Bombay, when dispatches
arrived from the Admiralty stating that a much
larger armament would be sent from England under
Admiral Keppel, and that the East Indian squadron
was merely required to meet the fleet from England
at a rendezvous, and act under Keppel's orders.
This news upset all Admiral Cornish's schemes, and
it must have caused him considerable disgust to find
his own ideas adopted, but given to another to
execute. He obeyed his orders by sailing to the
rendezvous appointed for both fleets, which was the
island of Rodriguez (or Diego Reyes), where he arrived
with his squadron in October, 1761, and where he
waited in vain for Keppel. It will not be incredible
to those who know the ways of officials, that the expedition announced to Admiral Cornish, and which
he was to meet at Diego Reyes, was never fitted out,
never sent, and never existed outside the dispatch
which was written to thwart and annoy the admiral at
Bombay. There can be very little doubt that if
Cornish had been allowed to carry out his own plan
the islands of Bourbon and Mauritius would have
been taken without difficulty.

As it was, the East Indian squadron waited for weeks and weeks at Diego Reyes, expecting the imaginary fleet under Admiral Keppel which never existed. This island at Rodriguez, ten miles long by four broad, is 330 miles east of Mauritius, and was discovered by the Portuguese navigator, Pedro Mascarenhas, in 1512. It was first visited by a French ship in 1638, and was fully described by Leguat in 1693. The roadstead is called Port Mathurin, and the Governor of Bourbon took possession of the island, and established a guard there under a superintendent in 1725. In 1756 an establishment was placed on the island for supplying Mauritius and Bourbon with turtles. Admiral Cornish found one Frenchman and his wife and half-a-dozen blacks left to collect turtle.

Young Rennell occupied most of his time, during his long stay at Rodriguez, in making a survey of Port Mathurin, which he finished early in December; and he also wandered over the island, which he described as very rocky and barren. This was the abode of that extraordinary bird called the "Solitaire," allied to, but distinct from, the Dodo. It is now extinct, but the Abbé Pingré believed that a few survived until 1761, though Rennell does not mention having seen one. The island and its strange products made a deep impression on his mind; and many years afterwards he described to his little grandson a species of very large spider belonging to Rodriguez, beautifully coloured, and which lived in the trees, carrying its web from tree to tree in the forests. He added : " I have caught some of them, and pulled out the thread with my fingers, which was as strong as common sewing thread. The orifice was visible, with a kind of raised ring round the edge of it." He also mentioned

the purslane (*Portulaca oleracea*) as common in waste
ground, " which we get in large quantities, and is an
excellent anti-scorbutic." Leguat* also refers to the
abundance of purslane in some places of the valleys.
Rennell saw the great land tortoises with a carapace four
and a half feet long, " and flesh like mutton, but more
delicate," says Leguat, which have since been exter-
minated. The sea-coast swarmed with turtle, " which,"
says Rennell, " make excellent soup; and we are supposed to have eaten 60,000 since our arrival; and
there is plenty of fish." The squadron was over
seven weeks at Rodriguez, and afterwards cruised off
Port Mathurin, waiting for Keppel's fleet. Admiral
Cornish waited until the stock of fresh provisions was
exhausted, and then made the best of his way to
Madras. During the voyage his squadron was visited
by a frightful outbreak of scurvy. The *America* lost
160 men in three weeks, and only had 32 out of the
sick list. Out of the *Grafton's* complement of 520
men, only 150 remained well; but young Rennell
never felt an hour's sickness.

Having completed his six years' service as a midshipman, Rennell was now eligible for promotion to
the rank of lieutenant, and his future career became
a very momentous question for him. The *Grafton* had
arrived at Madras in March, 1762. Captain Parker
foresaw difficulty in getting his young shipmate a
commission, well as he deserved promotion. He
offered to make interest with Lord Anson to obtain
him a lieutenancy, but also suggested that it might,
on the whole, be better for him to enter the sea
service of the East India Company. Meanwhile, the

* Page 70.

Grafton went to Trincomali, and the cruise gave young Rennell leisure to think the matter over. It was at this time that he received news of the death of one of his best and kindest friends, Miss Burrington, sister of the vicar of Chudleigh.

In April, 1763, the *Grafton* was again at Madras, and Captain Parker arranged that his young friend should be lent to one of the Company's men-of-war bound to the Philippine Islands, with the object of establishing new branches of trade with the natives of the intervening places. Rennell's duty was now to delineate the coasts and to draw charts. By this arrangement he would be able to acquire some experience of the Company's service, while he could return to the Royal Navy if he came to the conclusion that such a course would be preferable. Unfortunately, there is no account of this cruise to the eastern islands, which appears to have occupied about a year ; but there are five charts drawn by Rennell at this time, which were engraved by Dalrymple, and are now in the collection of the India Office.* They show that the Nicobar Islands, Quedah, and other places on the Malay Peninsula, the Straits of Malacca, and places on the north-west coast of Borneo were visited. Rennell described the voyage as a series of dangers, disappointments, and hardships. But his useful services recommended him to the authorities at Madras, and on his return he was offered the command of one of the Company's ships of war of fourteen guns, with a crew

---

* Bay of Camorta, Nicobar Isles 1762
Quedah 1762
Sambeelan Isles, Straits of Malacca 1762
Malacca ... ... July, 1762
Abai Harbour, N.W. Borneo.. `.. ... 1762

of a hundred men. As he was not discharged, and Captain Parker was absent, he was obliged to decline the offer.

All hope of advancement in the Navy for young officers without interest was put an end to by the peace of Fontainebleau, which was signed on the 10th of February, 1763. A final decision was come to in favour of the East India Company's service, and Rennell was discharged from the navy at Madras in July, 1763, after a service of seven and a half years. He at once received command of a ship of 200 tons, with a salary of £300 a year : ten times that of a midshipman. But his command was of very brief duration. On the 21st of October, 1763, a hurricane destroyed every ship in Madras Roads, not two Europeans being saved out of the crews of twelve large vessels. Providentially, young Rennell was on shore, but he lost everything he had in the world. He had, however, made numerous friends at Madras ; and among the warmest and most active was the Governor himself, Mr. Robert Palk,* who soon found employment for the youthful sailor where his services would be most useful. The home of the Palks, on Haldon Hill, almost overlooked the little town of Chudleigh, so that home feelings may have had something to do with Mr. Palk's steady friendship, which endured through life.

Rennell was appointed to the command of a small vessel, called the *Neptune*, belonging to a worthy

* Mr. Robert Palk, the Governor of Madras in 1763, had married Anne, sister of Mr. Henry Vansittart, the Governor of Bengal. He was created a baronet in 1782, and died in 1791. His son, Sir Lawrence Palk, Bart., was M.P. for Devon, and his great-grandson was created Lord Haldon in 1880.

Madras merchant, and he was recommended by the
Governor as a proper person to superintend the
landing of stores and the disembarkation of troops
for the siege of Madura, in the extreme south of the
Madras Presidency. At that time Madura was held
by Muhammad Isuf, who had formerly commanded
a native force in the Company's service. But he now
refused to pay any revenue, and a force was sent to
supersede him, which he resolutely opposed. The
siege of Madura, therefore, became necessary, and the
place was not taken until after a long resistance. On
the 16th of December, 1763, Rennell sailed in the
Neptune on the duty with which he had been entrusted, and which he performed to the satisfaction of
the Governor. The troops for the siege of Madura
were landed without accident, and Rennell was then
ordered to remain between Ceylon and the continent,
in charge of a fleet of small vessels, ready to land
reinforcements. It was at this time that he executed
surveys about Cape Calimir and the Pamben
Channel, and the strait between Ceylon and
Tinnevelli was named after the Governor—Palk
Strait. After the completion of this arduous work,
Captain Rennell returned to Fort St. George, and had
the gratification of receiving the thanks of the Madras
Government and a handsome present of money.
While commanding his little squadron, he held the
local rank of commodore.

"I then went to Bengal," wrote Rennell, "in my
owner's ship, where I met my worthy friend, Captain
Tinker, in command of the king's squadron there.
He called upon Mr. Vansittart, the Governor, and
procured me a commission as Surveyor-General of the
East India Company's dominions in Bengal. A few

days afterwards I had another commission sent me
as Probationer Engineer in the citadel erecting at
Calcutta, near Fort William. I must confess that I
was never more surprised in my life. I have since
found that I was indebted to one Mr. Topham, a
young gentleman who formerly served in the Navy in
the same capacity as myself, and who, having been
the most fortunate of the two in rising in the world,
thought it his duty to promote my interests as much
as possible." Rennell's commission as ensign in the
Bengal Engineers is dated April 9th, 1764. He was
twenty-one years of age.

Surely in this short paragraph there is as wonderful a story as is to be found in the Arabian Nights.
A discharged midshipman, in command of a small
coasting vessel, arrived at Calcutta, having lost all he
possessed by shipwreck, and forthwith he is appointed
Surveyor-General of Bengal, with a handsome salary
and brilliant prospects. But in reality everything
had been leading up to this unexpected consumma-
tion. Young Rennell had ever acted up to the motto
of Prince Henry the Navigator, *Talant de bien faire*:
he never spared pains and he strove to do well. On
such men Fortune seldom turns her back. Moreover,
he chose the line along which he would work in his
earliest youth, and never swerved from it. He would
be a geographer, and with that final aim he must first
be a surveyor. The little boy made a plan of Chudleigh before he went to sea. We have seen him
busily at work while the disaster was proceeding on
that fatal beach of St. Cast. During the voyage to
India, without help or encouragement, he never missed
an opportunity of surveying the ports in which the
America anchored. In happier circumstances he

surveyed and made a chart of Trincomali Harbour, with the warm approval of Captain Hyde Parker; so that he began to be known in the fleet, and even beyond the fleet, as a diligent and enthusiastic young surveyor. In that capacity he was lent to one of the Company's ships; and while he served on board her he was constantly employed in surveying and drawing charts. Finally, he had done very useful work in Palk Strait and the Pamben Channel; so that Governor Palk, of Madras, was able to testify to his brother-in-law, Mr. Vansittart, the Governor of Bengal, that the youthful captain of the *Neptune* was an expert and thoroughly reliable surveyor. His merits were well-known in the fleet, and Captain Tinker could conscientiously confirm the report from Madras. There could be no question as to young Rennell's qualifications; the only thing that could be brought against him was his extreme youth.

It may, however, well be doubted whether such testimonials would have sufficed if private and powerful interest had not been brought to bear in the right quarter; nor perhaps would that interest have been of sufficient weight without the testimonials. Rennell—fortunately for himself—had the faculty of making friends. Obliging, cheerful, and self-respecting, he was liked by his messmates. The truest and most enduring friendships are formed in a midshipman's berth; and on board the *Brilliant*, when midshipmen together, Rennell and Topham were sworn chums. Topham afterwards got an appointment in the Company's civil service, was advanced rapidly, and had indeed made his fortune. He was at Calcutta when the *Neptune* arrived, and high in the Governor's confidence. Mr. Vansittart was the first

British ruler in India who felt the importance of accurate surveys, and he was anxious to inaugurate some system for at least correcting and revising the received geography of Bengal. This was the moment for Mr. Topham to bring all his interest to bear for the benefit of his old messmate, and he did so with a success which aroused Rennell's surprise. The Surveyor-Generalship was obtained firstly through the *Talant de bien faire:* the persevering struggle to do well during several years; and secondly, through one of those friendships formed in a midshipman's berth, which are ever so lasting and so true.

Rennell had received an admirable training. The Navy has been the nursery not only of renowned seamen, but also of great generals, lawyers, divines, and men of science. There could be no better preparation for a geographer; and Rennell came direct from a midshipman's berth to the office of Surveyor-General of the Company's dominions in Bengal. He was a sailor; and the first of English geographers served upwards of seven years as a midshipman, and owed the grounding of his knowledge, and his capacity for work, to the sea. Many of his high qualities, as a scientific geographer, developed in after years, are directly traceable to his naval training.

CHAPTER III.

RENNELL'S survey of Bengal was the first, and it is very creditable to British administration that it should have been commenced within six years of the battle of Plassy and the acquisition of the country. After a residence at Fort William, and making the necessary preparations, which occupied a month, the young sailor took the field, commencing work on the Ganges, and carefully fixing points along its course. During this time, and throughout the autumn of 1764, he was living on board a very badly fitted-up budgerow. But the work of this first season was merely tentative, and meanwhile, the Governor had collected all the materials he could find. They were, however, of little value, and when Rennell returned to head-quarters, he made proposals for correcting the whole geography of Bengal.

During the first season Rennell acquitted himself to the full satisfaction of the Government; and on embarking for England, Mr. Vansittart wrote to him : " As the work you are now employed on will, I think, be of great use, so nothing in my power shall be wanting to put your services in such a light to the Company that they may give you the encouragement that your diligence deserves." His allowances were £900 a year, which, with other perquisites, made his income up to £1,000. As soon as he was in receipt of it, he

settled pensions on his mother and sister, and sent
home handsome presents to the Burrington boys—the
" dear children " of his early midshipman letters. On
the 14th of January, 1765, he got his promotion, and
became a lieutenant of engineers.

In the second season the scheme of operations for
a comprehensive survey was matured, and Rennell
completed a square of 22½ degrees of longitude and
3 of latitude, which brought him within sight of the
mighty chain of mountains separating the plains of
Bengal from Tibet. The name of Himalaya was then
unknown, and Rennell called them the " Tartarian
Mountains." He also completed the mapping of the
course of the Ganges within Bengal. Lord Clive
arrived at Calcutta on the 3rd of May, 1765, and in
October he ordered that another officer should be
appointed to the survey as Rennell's assistant. This
was Ensign W. Richards. Rennell wrote:—" I have
now company at all times; and luckily for me,
the gentleman proves a very agreeable and cheerful
companion."

Rennell fixed his head-quarters during the recess
season at Dacca, the old Muhammadan capital of
Bengal in the seventeenth century. Standing at the
junction of the river systems of the Ganges and
Brahmaputra, Dacca was a tolerably central position
for his work. On the east is the Megna, on the south
and south-west the Padma, or main stream of the
Ganges, and on the west the Jamma, or present
channel of the Brahmaputra. It is surrounded by a
perfect network of fluvial highways, the city itself
being on the north bank of the Buriganga, and extending four miles along the river-bank. Rennell built
himself a house at Dacca, and here he worked out the .

field operations and prepared his maps. He became attached to the place, and formed many lasting friendships. Among his Dacca friends were Mr. Cartier (afterwards Governor of Bengal) and Mr. Kelsall (the Company's resident), Colonel Claude Martin, and Dr. Russell. An intimate acquaintance also sprang up between Rennell and Mr. Hugh Inglis,* who was engaged in commercial pursuits at Dacca.

In the cold season of 1776 Rennell extended his operations to the frontier of Bhutan, where he met with a most serious accident, being so desperately wounded that his life was despaired of, and his constitution was permanently injured. The Sanashi Fakirs, part of a fanatical tribe, were in arms to the number of 800, while he was engaged in surveying Baar, a small province near the Bhutan frontier. They had taken and plundered a town within a few miles of the route of the surveyors, and Lieutenant Morrison was sent against them with ninety Sepoys. Morrison had been a midshipman on board the *Medway* when Rennell was in the *America*, and they went out to India in company. As soon as he heard that his old naval friend was on this service, Rennell at once set out to join him, and came up with him after he had defeated the Sanashis in a pitched battle. His detachment, being tired out, rested on the ground that night. Although Rennell was senior to Morrison,

---

* Hugh Inglis was born in Edinburgh, and was a commission merchant at Dacca for several years, while Mr. Cartier was resident. When Mr. Cartier went to Calcutta as second in Council in 1767, Mr. Inglis followed him, and continued his business on a more extensive scale. Returning to England, with a fortune, in 1775, he was elected E.I.C. Director in 1784, and was Chairman in 1812. He was created a baronet in 1801, and died in 1820, aged 77.

as his friend had been entrusted with the duty, he
chose to serve under him as a volunteer rather than
interfere with his command. Next morning they all
marched in search of the enemy. After a fatiguing
and tedious, movement, by which they hemmed in
the Fakirs between the forks of the Brahmaputra,
they found it necessary to reconnoitre a village in
their road, although they had no expectation of any
hostile force being there. But they were soon un-
deceived on finding themselves in front of two lines
of the enemy drawn up in the market-place. Their
escort of a few Horse rode off, and the enemy, with
drawn sabres, immediately surrounded them. Morrison
escaped unhurt. Richards received only a slight
wound, and fought his way out. The Sepoy adjutant
was badly wounded, but got off. Rennell's Armenian
assistant was killed. He himself was so completely
surrounded that he had little prospect of escaping.
His pistol flashed in the pan. He had only a short
sword, and with that he kept retreating backwards
until he thought he had few of the enemy behind
him. He then turned, and ran for it. One of the
Fakirs followed him a little way, but paid the price of
his life. The rest thought he was too badly wounded
to run far, but kept up a constant fire on him all the
time he was in sight. Rennell soon found himself
fainting through loss of blood, and the remainder of
the detachment coming up, he was put into a palan-
quin. Morrison then made an attack on the enemy,
and cut most of them to pieces.

Rennell was in deadly peril. He was deprived of
the use of both arms, and the loss of blood threatened
immediate death. One stroke of a sabre had cut his
right shoulder-blade through, and laid him open for

nearly a foot down the back, cutting through or wounding several of the ribs. He had, besides, a cut on his left elbow, a stab in the arm, and a deep cut over the hand, which permanently deprived him of the use of a forefinger. There were some other slighter wounds, and a large cut across the back of his coat was found when it was taken off. It was, fortunately, a thick regimental coat; but if he had happened to have been wearing his usual thin clothing, this cut would probably have terminated his existence.

There was no surgeon nearer than Dacca, so he had to be taken for three hundred miles in an open boat, which he had to conn himself while lying on his back, while the natives applied onions as a cataplasm to his wounds. He was six days in the boat, and when he arrived, and for many days afterwards, he was entirely given up by the surgeon. By very slow degrees he recovered the use of his limbs, and by the end of May his wounds were healed; but he never had the perfect use of his right arm again, nor the forefinger of his left hand, while the loss of blood permanently injured his constitution. "My companions," he wrote, "thought it almost miraculous that I have escaped so well; and I am very thankful that I am not entirely deprived of the use of my right arm, the provider of my daily bread."

Rennell's recovery was due to the affectionate care of Dr. Francis Russell, the station surgeon at Dacca, who passed the prime of his life there, and whose friendship for the young surveyor continued fresh and unbroken until death.* In consequence of

* Dr. Russell died at Bath on August 5th, 1791, aged 68, and Major Rennell wrote the inscription which was placed over his tomb in Walcot Church.

D

this serious accident, Lord Clive ordered that in
future the Surveyor-General should be attended by a
company of Sepoys.   During the year 1766 Lord Clive
kept Rennell very fully employed on his own affairs,
as well as on the public service.   He encouraged him
to complete the general survey and map of Bengal,
communicated to him all the materials that could be
found in the public offices, furnished him with a
proper establishment, and gave him all the assistance
in his power.   Finally, Lord Clive created for him the
office of Surveyor-General, with the rank of Captain of
Engineers.   Rennell prepared for the Governor a map
of Bengal, Bahar, and Orissa, and of the Mogul
Empire as far as Delhi, as well as a chart of the
Ganges, which were sent home for the use of the
historian Orme, who wrote such an admirable account
of the early military achievements of the English
in India.*

It was, perhaps, fortunate that Captain Rennell
was so fully occupied on detached duty during 1766,
or he might have made common cause with the insubordinate officers who struck for double *batta;* for
he seems to have felt strongly on the subject.

Double *batta* was first introduced after the battle
of Plassy by the Nawab Mir Jaffier, who granted it to
the English officers and soldiers for whose pay he was
liable, according to the Treaty.   Lord Clive, at the
time, warned the army that it was an indulgence on
the part of the Nawab which the Company could not
continue.   Accordingly, when the Nawab assigned

* The first part of Orme's " *History of the Military Transactions
of the British Nation in Indostan* " (1745–1760) appeared in 1763, in
three volumes.   Orme's " *Historical Fragments*," 4 vols. 4to., came out
1763–1805.

certain districts to the Company, to defray the expenses of the army, the directors issued orders that double *batta* should be abolished. These orders were repeated several times, but the remonstrances of the army had hitherto prevented the Governors in Council from carrying them out. The extreme unpopularity of the measure caused the orders to be evaded. When Lord Clive left England in 1764, this was one of the points most strongly pressed upon him by the Court of Directors.

Soon after his arrival at Calcutta, Lord Clive issued the order that double *batta* should cease on the 1st of January, 1766. The army was then in three brigades: the first under Sir R. Fletcher at Monghyr, the second under Colonel Smith at Allahabad, and the third under Sir Robert Barker at Bankipur. In reply to the remonstrances of the officers, the orders of the Company were stated, and the reduction of pay took place at the appointed time. Secret meetings of officers were then held in each brigade, at which a general resignation of commissions was proposed. Nearly two hundred commissions of captains and subalterns were collected, to be placed in the hands of the officers commanding the brigades on the 1st of June. They resolved to refuse the usual advance of pay for June, in order to avoid the charge of mutiny.

In March, 1766, Lord Clive and General Carnac set out from Calcutta, to regulate the collection of revenue at Murshidabad and Patna, and to form alliances against the Mahrattas. In April Sir R. Fletcher reported a rumour to Lord Clive that the commissions were going to be sent in, and the victor of Plassy replied: " Such a spirit must at all hazards be suppressed at its birth, and every officer that

D 2

resigns his commission must be dismissed the service."
He also sent to Madras for officers. On the 1st of
May officers of the third brigade sent in their commissions to Sir R. Barker, and forty-two officers of the
first brigade to Sir R. Fletcher. Lord Clive arrived at
Bankipur on the 15th, cashiered the ringleaders, but
allowed officers who had continued to do duty to
remain. Sir R. Fletcher was accused of being a
party to the conspiracy, tried by court-martial, and
cashiered, with six other officers, who were sent home.

The mutiny was quelled; and Lord Clive concluded
his period of office by a noble act of munificence. He
had received a legacy of £70,000 from Mir Jaffier,
which he lodged in the Company's treasury, in trust
to form a fund for the relief of disabled or decayed
European officers and soldiers, and for their widows.
His lordship finally left India in January, 1767.

Captain Rennell's feelings were, perhaps naturally,
with the officers. He wrote that about forty remained,
including himself, for that owing to being employed
on detached service, he had not the honour of being
in the secret. "It was," he added, "indeed a lucky
circumstance for me, for no doubt I should have been
carried away with the stream, and should have
entered into an association which has been attended
with disgrace to all those concerned in it." But
Rennell felt the reduction of pay rather bitterly, and
sympathised with the insubordinate officers. In one
of his letters home at this time, he wrote:—" In this
affair we discover the generosity of the Company to a
set of men who have conquered a territory equal in
extent to the kingdom of France, and this in a climate
that proves so prejudicial to European constitutions
that scarce one out of seventy men returns to his

native country. There is a passage in Rollin's
' Ancient History ' relating to the Carthaginians dis-
banding their mercenary troops after those troops had
preserved the State from ruin, and the reflection on it
perfectly suits the East India Company. Then,' says
Rollin, ' you discover the genius of a State composed
of merchants, who make a traffic of their fellow-
creatures.' " By this reduction the Surveyor-General
lost six rupees a day.

Mr. Henry Verelst, who succeeded Lord Clive as
Governor of Bengal in 1767, was a very good friend of
Captain Rennell ; and Mr. Cartier, who was second in
the Council, became Governor after the departure of
Mr. Verelst in 1768. Mr. Cartier had been Rennell's
intimate friend at Dacca, and was inclined to favour
him to the utmost. Indeed, for some time before
his own house was completed Rennell formed one of
Mr. Cartier's family.

In the season of 1767–68 the Surveyor-General
was at work in districts to the east of the Brahmaputra,
and also in Rangpur and in Rangamati, on the right
bank of the Bhagirathi. Here the Butanese drew up
an army to oppose his progress, and Rennell very
nearly fell into an ambuscade, but escaped with only
one man dangerously wounded. He described the
forests of Rangamati as chiefly inhabited by wild
buffaloes, elephants, and tigers. In the following years,
1768–69 and 1769–70, Rennell took his surveyors
farther to the eastward in the valley of the Brahmaputra—a wild country, infested by savage animals. On
one occasion the Surveyor-General was engaged with
his measurements on the verge of the jungle, when a
large leopard sprang out at him. He was so fortunate
as to kill the beast by thrusting his bayonet down its

throat, but not before five of his men had been
wounded, some of them very dangerously. Indeed,
the work of making a survey of Bengal was surrounded by difficulties and perils of all kinds, and
since his severe wounds Rennell had had frequent
attacks of ague and fever.

In March, 1771, the Surveyor-General was ordered
to take command of an expedition against a body of
men who made inroads into the northern provinces,
and levied large contributions. He marched three
hundred and twenty miles in fifteen days : a remarkable march for soldiers in such a climate, as he justly
remarked. The enemy, however, outmarched them, and
would never have been overtaken but by the help of
another detachment, which moved towards them from
the opposite direction. As it was, Rennell returned
successful, having completely secured the objects of his
enterprise. But it was at the cost of another violent
attack of fever, which nearly carried him off. The
preservation of his life was again due to Dr. Russell's
assiduous care.

The Surveyor-General, besides his house at Dacca,
had an office at Calcutta, and was there during part of
every year. He was always welcome at the house of
Mr. Cartier, and there he made the acquaintance of
the two Miss Thackerays, whose brother was in the
Civil Service, and Mr. Cartier's secretary. The marriage of James Rennell was so important an event in
his life, that it seems desirable to give some account
of his wife's family.

The Thackerays of Hampothwaite, on the Nidd, in
the West Riding of Yorkshire, had been small free-
holders for two hundred years, when two brothers,
Elias and Thomas Thackeray, were born in the last

decade of the seventeenth century. Elias became rector of Hawkswell. Thomas was a King's Scholar at Eton, Fellow of King's, a Master at Eton, and in 1746 he became Head Master of Harrow; and he was also Archdeacon of Surrey. He married Ann, daughter of John Woodward, of Butler's Marston, in Warwickshire, and when he died, in 1760, he left a widow and fifteen children. The archdeacon's widow continued to live at Harrow with her unmarried daughters until her death, at the age of 89, in 1797. The eldest son, Elias, was Vice-Provost of Eton, and died unmarried in 1781. The second, John, who was chaplain at St. Helena, also died childless in 1770. Thomas, the third, was a surgeon at Cambridge, and, like his father, had fifteen children. His grand-daughter, Mrs. Bayne, wrote a history of the family. Frederick, the fourth son, a physician at Windsor, had several children, among whom George was Provost of King's, and Frederick Rennell was a general R.E. and C.B. Joseph, the fifth son, was forty-three years Receiver-General of Customs, and lived unmarried with his mother and sisters at Harrow. William Makepeace Thackeray, grandfather of the great novelist, was the youngest son. Born at Harrow in 1749, he went out to India in the Bengal Civil Service in 1765. There he became secretary to the Governor, Mr. Cartier, and in 1776 married Amelia, daughter of Colonel Richmond Webb, by whom he had nine children.*

---

* William, the eldest, was a Member of Council at Madras with Sir J. Munro, and died in 1823.

Emily married John T. Shakespear, D.C.L., and was the mother of Sir Richmond Shakespear.

Richmond went out to India, and died in 1816. By his wife, Anne

There were also eight daughters, and it was arranged that two of them—Jane and Henrietta—should go out to India, and join their brother William at Calcutta. Jane was born at Haydon, in October, 1739, and Henrietta, who was seventeen years younger, in 1756. Jane was not beautiful, but she was a sweet, sensible, unaffected woman, and much beloved. Henrietta was considered a beauty. James Rennell became attached to Jane Thackeray, and they were engaged for nearly a year. The marriage took place at Mr. Cartier's house at Calcutta, on October 15th, 1772, and five days afterwards the Surveyor-General and his wife set out for Dacca. In November Rennell announced his marriage to Mr. Burrington. "I have," he wrote, "every prospect of felicity in my present state, and want nothing more than to be settled in my native country." He was just thirty-one, his wife three years older. Soon afterwards, Henrietta came to visit her sister at Dacca, and almost immediately became engaged to Mr. Harris, the Resident. Mr. and Mrs. Harris went home in the same ship with Mr. Cartier, the retiring Governor, who was relieved by Warren Hastings in 1772. In July, 1773, a little girl named Jane was born to the Rennells, but she died one year and six days afterwards. A little silver copy

Becker, he had a son: William Makepeace Thackeray, the great novelist, born at Calcutta.

Charlotte married John Ritchie, and was mother of William Ritchie, Legal Member of the Supreme Council at Calcutta, who died in 1862, leaving, with others, Augusta, wife of Douglas W. Freshfield, Secretary of the Royal Geographical Society, 1882–1894.

Francis, in holy orders, author of a "Life of Lord Chatham." He married Anne, daughter of J. Shakespear, and had a son, Colonel Edward T. Thackeray, R.E., V.C., and four others: Augusta, Thomas, St. John, and Charles.

of her tomb at Dacca was made for the parents, with the name and date inscribed on it, which is still preserved.

The Surveyor-General had now nearly completed his work in the field, and for the next few years he was mainly occupied in arranging and collating his materials, and drawing the maps for the "Bengal Atlas," which was to be in fourteen sheets. Much of his manuscript work is still preserved at the India Office. The "Map of the Denospur and Rangegunge Rivers," surveyed in 1767 by Rennell and Richards, is on a scale of three miles to an inch. There are also a plan of the British factory at Dacca on a large scale, a reduced general map of the Megna and other rivers, a large scale survey of the River Megna, the eastern branch of the lower creek, the Mandapur Creek in three parts, the Ganges fifteen miles above and below Colgong, the River Brahmaputra in five parts, and maps of Rangpur, Rangamati, and Kuch Bahar. But these are only a small remnant of the work ot Rennell; and it was in the preparation of the "Bengal Atlas" from his numerous large scale maps which were the results of his field work, that the Surveyor-General was occupied during the last years of his residence in India. The "Bengal Atlas" was first published, in one volume folio, in November, 1779, and the second edition appeared in 1781. It was a work of the first importance both for strategical and administrative purposes, and is a lasting monument of the ability and perseverance of the young Surveyor-General. He executed it between his twenty-first and his thirty-sixth years. Each map was dedicated to friends with or under whom Rennell had served; and it is interesting to see who were those selected by the Surveyor-

General for this distinction. One map is to the
memory of Lord Clive, and others to Governors
Verelst, Cartier, and Warren Hastings. Four maps
are dedicated to officers who commanded the troops
in India in Rennell's time: namely, Sir Hector Monro,
Sir R. Barker, General R. Smith, and Captain J. Jones;
to two officers who assisted him—General Caillaud and
Major Carnac; and one to Mr. Broughton Rouse, who
translated for him several passages from the Ayin
Akbari. The rest are inscribed to the long-tried
friends at Dacca, Mr. Kelsall, the Resident, Sir Hugh
Inglis, and Dr. Francis Russell.

Rennell's happiness had been ensured by his
most auspicious marriage. In those days there were
no hills, no refreshing sanatoria on the slopes of the
Himalayas, to which the Europeans whose duties kept
them down in the hot and steaming plains could
resort for the restoration of their exhausted powers.
The best change of air that Rennell could find for
himself and his wife was on the long and narrow strip
of coast-line, backed by low ranges of hills, which
forms the province of Chittagong, about 165 miles
long by fifteen broad. The Sitakund range of hills
contains the sacred peak of Chandramath, 1,155 feet
above the sea, and beneath it is the capital of the
province, called Islamabad, which was ceded to the
English in 1760. The European houses are each on
small steep hills, but the place is now considered
unhealthy.

It was to Islamabad that Major * and Mrs. Rennell
resorted to recruit their strength after a long residence
at Dacca. Writing thence to Mr. Burrington, in 1776,
he said that Chittagong was considered the Montpellier

* Rennell received the rank of major on April 5th, 1776.

of Bengal, and described it as a hilly country bordering on the sea. In the same letter he dwelt upon his great happiness. " I cannot help repeating," he wrote, " how supremely happy I am in possessing such a woman as Mrs. Rennell. Temper is, I believe, the basis of love and friendship. Neither the wittiest nor the wisest bear away the palm of happiness and content, therefore I conclude it depends on temper. At the same time—*cæteris paribus*—I believe the wisest are the happiest, as it is these alone that make allowances for the frailties and imbecilities of human nature."

The survey of Bengal having been completed and the sheets nearly ready for publication, Major Rennell made preparations for returning to his native land; for his wounds had permanently injured his constitution, and made it impossible for him to prolong his residence within the tropics. His application for a pension was met in a most liberal spirit by Warren Hastings, who valued his services very highly. His Government settled a handsome pension of £600 a year on the retiring Surveyor-General, subject to confirmation by the Court of Directors.

A passage was taken in the *Ashburnham* (Captain Waghorn), which was to leave Calcutta in March, 1777, and the Rennells finally left Dacca on the 2nd of February. The major's brother-in-law, William M. Thackeray, had sailed in the *Triton* with his wife in the previous December. On arriving in England, he bought a house at Hadley, near Barnet, where he continued to reside, bringing up a numerous family, until his death, in 1813. Rennell would find many changes when he arrived in England after so long an absence. His mother had died in 1776. His sister

Sarah had made an imprudent marriage with a man
named Edwards, and henceforward relied much on
pecuniary help from her brother, which was freely and
generously given. The Burringtons were alive and
well, and the boys were grown up. Tom had entered
the army. Robert, by Rennell's advice, had joined the
sea service of the East India Company, and was at
Calcutta as a midshipman in 1772, but, unluckily, was
not able to get leave to pay his old friend a visit at
Dacca. Gilbert was at Oxford. Major Rennell also
had to look forward to making the acquaintance of
the numerous members of his wife's family.

The *Ashburnham*, with the Rennells on board,
sailed from Calcutta in March, 1777, and arrived at
St. Helena, after a prosperous voyage. Mrs. Rennell
was so near her confinement that they decided to
remain on the island, where their daughter Jane was
born, on the 12th of October, 1777. They took their
passage home from St. Helena in the *Hector* (Captain
Williams), and had a tedious voyage of eleven weeks.
In January, 1778, a violent storm was encountered,
and the passengers were in great discomfort for several
days. All the live stock and dry provisions were lost,
and the ship was almost full of water; even their
infant was half drowned in the upper cabin. At
length, however, the long voyage home was at an
end. On the 12th of February, 1778, Major and
Mrs. Rennell landed safely at Portsmouth with their
little daughter, and set out for old Mrs. Thackeray's
house at Harrow the same evening. She was living
there with her bachelor son, Thomas, the Commissioner of Customs, and five unmarried daughters:
Decima, Theodosia, Alethea, Frances, and Martha,
who eventually married Mr. Evans, the curate. Mrs.

Rennell was to pay a long visit to her mother and sisters, while her husband recovered and recruited himself after the long voyage, occasionally going to London on business. He had temporary lodgings in Surrey Street, Strand, and afterwards in Oxford Street, near the corner of Orchard Street.

There was a doubt whether the Court of Directors would confirm the pension granted to the Surveyor-General by Warren Hastings. He visited the Directors, and also saw Lord Sandwich, from whose friendly interposition he expected more than from the spontaneous action of the Directors themselves. His friend Mr. Palk also applied to the Directors on his behalf, and gave them some testimonials of his services on the coast of Coromandel. In spite of Rennell's great services, the Directors reduced the pension granted by Warren Hastings from £600 to £400 a year; but two years afterwards the Court became thoroughly ashamed of their conduct, and granted the full pension recommended by the Government of India in November, 1781. Their parsimony is also far from creditable as regards the " Bengal Atlas." Instead of publishing it at their own expense, they allowed it to be brought out by a subscription of the Company's servants in India, and they would only advance the small sum of £150 to Major Rennell to enable him to proceed with the engraving of the maps, on condition that he gave them a bond to repay them in eighteen months.

It was on the 23rd of March, 1778, that Rennell set out to visit his old friends and the home of his childhood in Devonshire. He stopped at Bath for a few days, to see his friend Dr. Francis Russell, who had saved his life at Dacca by his tender and affectionate care. But he was at Chudleigh during the greater part of April, where he was warmly welcomed by old Mr. and Mrs. Burrington. Many an old haunt must have been re-visited, and many a long-forgotten incident brought back to his memory. But the greatest pleasure to Rennell must have been the society of his long-tried counsellor and guardian. In 1769 Mr. Burrington had been the means of recovering the estate of Waddon, after a lawsuit, and Rennell had written : " I am particularly pleased when I reflect how long it had been in the family." Waddon, his birthplace and the home of his childhood, was now his own. For this and many other acts of kindness, his gratitude was due to Mr. Burrington. He strove to repay it by the interest he took in the boys. Tom was in the army. He obtained the post of second mate in an Indiaman for Robert, in March, 1781. Gilbert was at Oxford, meditating the publication of a translation of Pindar—a project warmly encouraged by Major Rennell. In 1782 Gilbert became tutor to the sons of Sir Guy Carleton, who was afterwards created Lord Dorchester.

During the summer of 1778 Major and Mrs. Rennell were paying visits to Thackeray relations. In July they were staying with Mr. and Mrs. Harris (Henrietta Thackeray) at their house, called " The Vineyards," in Great Baddow, near Chelmsford. The Harrises lived in a very expensive style, and Mr. Harris drove a four-in-hand, so that when he died, in 1790, he left his property much embarrassed. The Rennells also visited Dr. Boscawen at Quendon, near Saffron Walden, Dr. Thackeray at Cambridge, and Mr. William Thackeray at Hadley, near Barnet. On returning to London, they took a house at 18,

Charles Street, Cavendish Square, in which their
eldest son, Thomas Thackeray Rennell, was born, on
the 22nd of May, 1779. Their second son, William,
was born on January 22nd, 1781. Early in 1781
they moved to the house which was occupied by
Major Rennell during the rest of his life, No. 23,
Suffolk Street (since called Nassau Street), near the
Middlesex Hospital. Here he was close to the house
of Sir Joseph Banks, the President of the Royal
Society, in Soho Square, who became an intimate
friend, and near many other agreeable acquaintances
in Wimpole, Harley, and Mortimer Streets.

Major Rennell suffered from very long and painful
illnesses in 1781 and 1782, the consequences of his
Indian service ; and he was depressed by the aspect
of public affairs and the wretched mismanagement of
the American revolutionary war. He wrote : " I hear
of nothing but misery and want among the lower
orders (the bulk of the people), and yet we are said
to be in a flourishing condition. To hear my Lord
North declare it, after exhausting his country, is too
much for my patience."

Rennell's last letter to the Rev. Gilbert Burrington
was dated January 18th, 1785. In that year he lost
his old and faithful friend, his constant correspondent
and agent since he left Chudleigh, a boy of fourteen,
nearly thirty years before. He must have felt this
loss very deeply, for the exchange of letters had been
continuous, and never so frequent as since Rennell
returned to England. Mr. Burrington was succeeded
in the vicarage of Chudleigh by his son Gilbert, who
held the living until 1841, when he died, at the age
of eighty-six.

As soon as the " Bengal Atlas " was published, Major

Rennell commenced his first purely geographical work : the Map of Hindostan, with a memoir. In 1782 he wrote to Mr. Burrington : " I have another geographical work in hand : a map of Hindostan, accompanied by about one hundred and ninety pages of letterpress. It is a work much wanted at this time. The map has just been a twelvemonth in the engraver's hands, and my illness has not hastened it." This work marks the time when Rennell ceased to be merely a surveyor and draughtsman, and became a geographer in the more extended sense. This, therefore, seems to be the place at which we should review the progress of geography as a science, and its position when Major Rennell began to take his place, in 1782, as the foremost geographer of the age.

CHAPTER IV.

MAJOR RENNELL was the leading geographer in England, if not in Europe, for a period of fifty years: from 1780 to 1830 ; while the influence of his example and of his methods has continued to be felt down to our own time. He inherited the accumulated stores of knowledge which were accessible in his day, and the scientific way of treating geographical questions which had been commenced by his French predecessors. The system of observing and of surveying in his time had been, to a great extent, elaborated by himself. It is necessary, before discussing his literary work, to remind ourselves of the character and extent of Rennell's geographical inheritance; for it is from that point of view that we shall have to consider the great progress that he was instrumental in securing for the mother of all the sciences.

His and our greatest inheritance was from the intellectual labours of the ancient Greeks; although he was dependent on English and French translations for his knowledge of them. As the ability of the Grecian race is higher by several grades than that of any people that has appeared since, we must ever look to it for our models in the exposition of geographical, as of every other branch of human knowledge. It was held by Sir Henry Maine that, in an intellectual

E

sense, nothing moves in this Western world that is not Greek in its origin; and it is, therefore, from the great masters of description and of philosophy in its wide scientific sense that modern geographers have received their most indispensable lessons. Herodotus, in his admirable descriptions to illustrate his history, furnished the earliest system of geography; Xenophon, as a practical traveller and an accurate observer, was a model for careful route surveyors; and Thucydides was undoubtedly the historian who most fully appreciated the importance of geography in the illustration of his own science. His lucid and masterly descriptions of particular countries, and his admirable treatment of topographical details, have never been surpassed. The ancient authors who afterwards devoted their attention more exclusively to geography, such as Strabo and Ptolemy, have, in another way, had an equal influence on modern systems and modern thought.

Strabo, in his minute investigation of the geography of that part of the world that was known in his time, brought a vigorous critical mind to bear on the facts which he had collected with extraordinary patience and industry. It is true that he was sometimes hypercritical, and, in rejecting information on insufficient grounds, he was led to wrong conclusions on several important points : as, for instance, in the position he gave to Ireland, and in his rejection of the prolongation of the peninsula of Brittany, as well as in some of his latitudes. But his systematic examination of evidence, and his methodical treatment of the materials within his reach, render Strabo a model who must always influence, directly or indirectly, the minds of modern geographers. Ptolemy, being a far more important author, has left a much deeper impression on modern thought. Indeed, his influence was almost paramount until the days of Ortelius and Mercator, and was felt down to within a century of the time of Rennell in certain departments of geographical work. We owe to Ptolemy all our knowledge of Eratosthenes, Hipparchus, and Marinus of Tyre; and his critical estimates of his authorities, the reasons he gives for the acceptance or rejection of their conclusions, and his own system of astronomical geography, have justly secured to him the highest place among the ancient writers on our subject. These great names establish the immense antiquity of geographical science—the first of all the sciences, and indispensable to some of the most important as a basis from which to work. The necessity for acquiring a knowledge of the ancient systems, and the usefulness of identifying the places and routes mentioned and described by the Greek authors, as forming one leading branch of his science, were ideas that were deeply impressed on Rennell's mind. He not only studied the principal geographical writers of antiquity with the aid of translations, but extended his reading to the works of every minor author that was accessible to him. Indeed, the great works of the ancients formed not only a valuable, but a very fruitful inheritance to the enlightened student, who gave its true value to comparative geography.

The Arabs were only transmitters of knowledge ; they added little to the stock accumulated by the superior race whose provinces they overran, while the injury they did to mankind by the destruction of the library at Alexandria is irremediable. The more enlightened successors of the first Muslim fanatics.

E 2

did good service by reproducing the work of Ptolemy.
There are also later Arabian geographers whose works
are valuable in the study of routes in Africa, Arabia
and Syria, Persia, and Central Asia; but they were not
all accessible to Major Rennell. He was, however,
acquainted with Edrisi, Ibn Haukal, Abulfeda, Alfra-
ganus, Benjamin of Tudela, and Ibn Batuta.

The Middle Ages left no valuable legacy to modern
geography. All was thick darkness, with here and
there a bright star, such as Roger Bacon and Sacrobosco. The Crusades gave rise to some imperfect
acquaintance with the East; there was some intercourse with distant countries through missions sent
at long intervals, and there were vague reports of
voyages. But even the little knowledge that did find
its way to the minds of the men who could read,
through these channels, was not appreciated or understood. Carpini, Rubruquis, and Marco Polo were the
heralds of the approaching dawn.

From the period of the Renaissance a great wealth
of geographical knowledge was inherited by posterity;
but in the days of Rennell much less of this knowledge was accessible than at the present time. Valuable maps and books which were then rare and costly
have now been brought within easy reach of students
by facsimiles and new editions. A great mass of
material was hidden away, through carelessness or
from political motives, more especially by the Spanish
Government, which has since been brought to light.
Nor do we even now know the full extent of our
geographical inheritance from the age of the Renaissance. Still, a vast mass of information, such as
it would occupy a student several years to become
completely acquainted with, was accessible to Major

Rennell. He was an indefatigable reader, and every work of importance was subjected by him to more than one perusal.

The first work of awakening civilisation in the fifteenth century, so far as geographical science is concerned, was the publication of editions of Ptolemy, each edition embodying maps and notes intended to represent and describe the latest discoveries. These editions were the work of learned scholars and professors. But there appeared at the same time those beautiful *portolani*, the work of Italian and Catalan cartographers from the information of sailors, which astonish us by the comparative accuracy of their delineations, and by the extent of the knowledge they represent. While the errors of Ptolemy are retained by the learned, we find them corrected or discarded by practical draughtsmen. Among these Catalan and Italian sailors the mariner's compass first came into use, and they first organised and led the Portuguese fleets on their career of glory. But it was Prince Henry the Navigator—" the immortal Prince Henry," as Rennell calls him—who commenced, by his genius and devotion, the period of continuous discovery. Stationed on the promontory of Sagres, occupied in training his courtiers and attendants to the work of exploring navigators, he sent forth expedition after expedition to discover the west coast of Africa; and so admirably had the prince organised his work, that on his death his own firm hand was never missed from the helm, and discovery proceeded until Bartolomeu Diaz rounded the Cape of Good Hope, and Vasco da Gama reached the west coast of India. The Portuguese made known during the first half of the sixteenth century the east coast of Africa,

70 DISCOVERY OF AMERICA.

Madagascar and the Mascarene Islands, the Eastern
Archipelago, the coasts of Malaya, China, and Japan,
and sent embassies into the interior of Abyssinia,
besides exploring the coast of Brazil. All these high
achievements were first recorded in the reports and
memoirs of the principal leaders of expeditions, and
then by the diligence of Galvano and Damian de Goes,
the masterly pen of Barros, and the genius of Camoens.
Thus the exploits and geographical services of the
little hero-nation was a part of the world's history,
and eventually became familiar to the students of
other countries.

The genius of Columbus opened a still wider field
for geographical discovery,* and the existence of a
new continent, believed to be almost as large as the
world known to the ancients, was explored in an
astonishingly short space of time. Fifty years after
Columbus landed on the islet of Guanahani, the whole
of the coasts of South America had been discovered,
as well as Mexico and Peru, and the western side of
North America, while Magellan had passed through
his straits and crossed the Pacific Ocean, and Sebastian
del Cano had circumnavigated the globe. Then
followed the construction of maps, many of which remained hidden until our own day, and the composition
of narratives and histories, some of the most precious
of which were kept secret and lost, or remained un-
printed until long after the time of Rennell.

The famous pilot, Juan de la Cosa, returning from
his third voyage across the Atlantic, established himself at the port of Santa Maria, in the Bay of Cadiz,

* See Rennell's remarks on the genius of Columbus and on the
globe of Martin Behaim, in his "Herodotus" II., p. 366.

and drew a great map of the world on an ox-hide, in
which he combined, with the discoveries of his
companions, Columbus and Hojeda, the work of the
Portuguese and of the English under John Cabot.
This most interesting and valuable document remained
hidden, and was only made known during living
memory. The beautiful Cantino map, drawn a few
years afterwards by a Portuguese draughtsman at
Lisbon, and chiefly intended to illustrate the discoveries of Corte Real, was recently recovered from a
butcher's shop at Modena, and published in facsimile
by Mr. Harrisse still more recently. It was the same
with the map of Sebastian Cabot, now at Paris, and
many others of like importance. They were all unknown in the last century. Others, including the
charts of Columbus himself, are irremediably lost.
Narratives and histories were, to some extent, equally
inaccessible. Robertson, in spite of all his researches,
never saw the letters or journal of the illustrious
Genoese, nor the invaluable history of Las Casas.

Still, even without the additional material that is
within the reach of students of the present generation,
there was tolerably complete information respecting
the history of geographical discovery in the New
World, and the progress of cartography, within the
reach of Principal Robertson of Edinburgh, and other
scholars whose acquaintance was enjoyed by Major
Rennell. The works of Peter Martyr, Oviedo, and
Gomara were known to them, and English translations
had been made of Cieza de Leon, of Acosta by
Grimeston, of Garcilasso de la Vega by Rycaut, and of
Herrera by Captain Steevens. The great collections
of Eden, Ramusio, Hakluyt, and Purchas, as well as
the later publications of Harris, Astley, and Churchill,

were known and highly esteemed. Eden and his
Elizabethan compeers not only gave English versions
of the letters of Peter Martyr and the chronicle of
Gomara, but also embodied the earlier French, English,
and Dutch efforts at discovery. Within these portly
tomes are the Portuguese exploits in India and further
east, our own early Arctic voyages, and the first
voyages of the East India Company. The travels of
Chardin, Tavernier, Herbert, Hanway, Pietro della
Valle, and others in Persia; the works of Petis de la
Croix, D'Herbelot, De Guignes, Du Halde, Georgi
Kœmpfer, Thevenot, Niebuhr, the Indian travels of
Bernier, Tavernier, and Sir Thomas Roe, were all well
known to Rennell; and he had also profited from
various translations of native works. Through
Hakluyt, he was made acquainted with the account of
Africa by Leo Africanus.

Although Wright was the first to explain the
principle of Mercator's projection, and to calculate a
table of meridional parts, pre-eminence in the cartographic art continued to belong to the Netherlands
and Germany, and afterwards to France, throughout
the seventeenth and indeed the eighteenth centuries.
We borrowed the Waeghaners, or volumes of sea
charts, from the Dutch; and Henry Hexham published
a magnificent translated edition of the atlas of
Hondius, with many of the plates engraved by
Mercator and his sons. Our own map-makers, including Moll and others, were very inferior, as regards
both knowledge and cartographic execution, to their
confrères in Amsterdam and Paris, although there
was a marked improvement later in the eighteenth
century, as is manifest in the engraving of the
" Bengal Atlas." Closely as he studied the history of

his science, Rennell never ceased to be a sailor and to watch all hydrographic work with the greatest interest. The geography of the sea was a branch of his subject which was quite in its infancy before Rennell himself and his indefatigable friend, Alexander Dalrymple, devoted their attention to it. Grenville Collins and a few others had made surveys on the English coasts, and their results had been published as a private speculation. Similar work had been done on the coasts of India, as will be seen in the next chapter; but as yet there was no systematic plan of operations with regard to marine surveys.

Besides the study of the history of geographical discovery, which he continued with such assiduity until he became a perfect master of the knowledge that was accessible in his time, Major Rennell watched all contemporary efforts with the keenest interest, both for the examination of inland regions and for the discovery and survey of coasts and islands. His friend Dalrymple's translations made him acquainted with the earlier work of Spanish navigators in the Pacific, and no one who was not on board—not Dr. Hawksworth himself—had a more intimate knowledge of the more recent voyages of circumnavigation by Anson and Byron, Wallis and Carteret, Bourganville and Entrecasteaux, and especially of the voyages of Captain Cook. The voyage of Anson had a peculiar interest for Major Rennell, because his old captain, Hyde Parker, was a master's mate on board the *Centurion;* and his intimacy with its commander gave him a similar interest in the voyage of Wallis.

Rennell's special knowledge enabled him to throw light on the authorship of the history of Anson's voyage, which bears the name of Mr. Richard Walter, the chaplain, whose widow claimed the work as that of her husband. The schoolmaster, Mr. Pascoe Thomas, also published a journal of the voyage of the *Centurion* in 1745. But the real author of the larger work attributed to the chaplain was an engineer officer on board, named Robins. Major Rennell observed, in a letter to a friend:* " I forgot to say, in defence of Anson's voyage, that a second volume, containing the nautical observations, was written and approved by Lord Anson. But Colonel Robins, being hurried off to India as Engineer-General, took the manuscript with him to revise and correct, very contrary to Anson's desire. Robins died not long after, àt Fort St. David, and the manuscript could never be found."

By his thorough mastery of geographical literature, both as regards the history of the progress of discovery and of cartography in the past, and the details of all current events and proceedings, Rennell made himself the highest authority in Europe in a very few years after his return from India, his fame as an able and intrepid explorer and surveyor having preceded him. He also took up the thread of the scientific treatment of geography where it had been dropped by his illustrious French predecessors. In earlier times the critical and philosophic method of treatment had not been brought to bear either on the preparation of maps or on the study of geographical questions. Information was recorded, additions were placed on maps as new discoveries were made known, but with less of that critical weighing of evidence which was necessary before geography could be restored to the position

a

* Barrow's " Life of Anson," p. 7.

of a progressive science: a position which it undoubtedly held in antiquity, in the days of Strabo and of Ptolemy. Such a spirit is partially visible in the work of Ortelius and Mercator; but after their time the map-makers, especially in England, became mere uncritical compilers. If they had worked with knowledge and intelligence, Baffin's Bay would never have disappeared, and the peninsula of California on Elizabethan maps would never have been converted into an island on those of Queen Anne. Yet Major Rennell has a good word for Moll, the English map-maker of Queen Anne's time. He refers to him as a cartographer " whose works we are too apt to contemn."

Even in France, during the seventeenth century, the leading geographers show no very great advance. The foremost in that period was Nicolas Sanson, who was born at Abbeville, of one of the most distinguished families of Ponthieu, in 1600, and who died at Paris in 1667. He wrote Latin dissertations on geography, began a map of Gaul at a very early age, and was appointed geographer to the king in 1627. Sanson produced a map of France in ten folio sheets, a map of the course of the Rhine, and a large map of Africa in 1656. His eldest son, Nicolas, gallantly rescued the Chancellor Seguier from the fury of the people during the Fronde, and escorted him, sword in hand, until, at the Pont Neuf, he received a shot from a musket, and died next day. But the two other sons had longer lives. Their names were Adrien and Guillaume Sanson. Both were geographers to the king, and they revised and re-edited many of their father's works. The youngest survived until 1708. In the productions of the elder

Sanson there are certainly signs of the spirit which actuated Ortelius and Mercator, and which is wanting in his English contemporaries. At all events, he and his work form the dividing line between the uncritical compilers of an earlier time and the period of scientific treatment which was at hand. There was an epoch, lasting several centuries, when nothing was done by geographers beyond the slow collection of materials and the study of Ptolemy's astronomical geography, distilled through such works as Sacrobosco's "De Sphærâ" and Pierre D'Ailly. Then there came an exciting period of discovery, when maps were made mainly for practical use on board ship, or compiled, with little or no discrimination, for more general use, and when great attention was given to instruments and other apparatus necessary for navigation. This wonderful and romantic age—the age of the Renaissance—bore rich fruit in its geographical and in other aspects. Its far-reaching results were illustrated by the great map drawn by Mercator for the Duke of Cleves, and later by the map (probably by Wright) which was used to accompany some of the copies of Hakluyt's voyages, and which was described by Shakespeare as "the New Map with the augmentation of the Indies." Sanson the elder was born in the year that this map appeared. Nothing better was produced in England, of purely English birth, during the following century. But Sanson and his compeers certainly made some advances both in the art of cartography and in the science of geography, and prepared the way for the more judicious and critical methods which had for their chief exponents, in the first half of the eighteenth century, those eminent successors of the Sansons, Delisle and D'Anville.

The father of the first of these great geographers was Claude Delisle, who was born at Vaucouleurs, near Toul, in 1644. He studied at the University of Pont à Mousson, and eventually went to Paris, where he gave lessons in history. Delisle *père* was the author of an introduction to geography and of a historical work on the Kingdom of Siam (1684), which reminds us of the connection attempted to be formed between France and that State as far back as the time of Louis XV., and even earlier. Guillaume Delisle *fils*, the great geographer, was born at Paris in 1675, was educated by his father, and became first geographer to the king. Young Delisle, like Rennell, was enthusiastically fond of geography from his childhood. At the age of nine he drew maps to illustrate ancient history; and when quite a young man he conceived a plan of re-forming and re-constructing the system of geography, which he matured in his twenty-fifth year. His map of Europe appeared in 1700, and two years afterwards he was admitted to the Academy in recognition of his great merit as a cartographer. His industry was on a par with his ability, and he published upwards of a hundred maps. Guillaume Delisle died in January, 1726. Delisle had been devoted to his science since boyhood; he was a man of untiring industry, and at the same time an original thinker. He finally cast aside all Ptolemaic trammels, and struck out a line for himself, seeking out information from every possible source, carefully combining and harmonising his materials, and making sure that there was good reason for every change he made and every position he adopted from older work. He was the first to restore geography to the position of a true science, by the adoption of scientific modes of thought and accurate reasonings in the treatment of all the geographical data, new or old, that came before him. Delisle and his successor may justly be considered, in a scientific sense, as the founders of modern geography.

D'Anville may, therefore, be looked upon as a disciple of Delisle; and this illustrious man of science carried his predecessor's exact methods to a higher perfection, while his long life enabled him to achieve great results. Jean Baptiste Bourguignon D'Anville was born at Paris in 1697, and was the son of Hubert Bourguignon by Charlotte Vaugon. When he was twelve years of age a map accidentally fell into his hands which interested him, and gave him a turn for geography. As a boy, he drew maps of the countries mentioned by the Latin writers he was studying, and this taste for drawing maps became a passion. He was a great reader; but he read poetry and history only to study the positions of the places that were mentioned. His whole soul was wrapped up in his love for geography. He loved it, not only because it helped to satisfy his devouring curiosity, but chiefly as a science to be honoured, and to which all the ability he possessed might be freely given. Before D'Anville had reached the age of twenty-two he was appointed geographer to the King of France. In the conversation of learned men he improved his habit of judicious criticism, and acquired scientific methods of thought which specially fitted him for his great mission; for it surely was a great mission to

examine with critical and discerning eye the whole
system of the geography of that day: to go over
the map of each region, judicially examining all
the data, rejecting some, modifying some, accepting
others, all on strictly scientific principles. It has
been thought that D'Anville was sometimes too
strict in erasing materials which did not seem to
him to be based on sufficient authority. On his
map of Africa a clean sweep was made of many of
the numerous names coming down, in an uncertain
way, from Edrisi and Leo Africanus, and in part
from Pigafetta and the Portuguese, and of all the
uncertain details. If this was a fault, it was a fault
in the right direction, and left nothing to confuse
or mislead the future discoverer. D'Anville published upwards of 210 maps and charts, and made
an immense collection of geographical documents,
which were acquired by the French Government in
1779. One of the most important studies of
D'Anville was undertaken to settle the length of
the measurements of the ancients. His " Geographic
Ancienne," in three volumes, came out between
1768 and 1782; and his maps, " Orbis veteribus
notus " and " Orbis Romanus," previously. D'Anville
died, at the age of eighty-five, in January, 1782: the
very year that Rennell commenced the second part
of his geographical life as an author, the first part
having been devoted to marine surveying, and to
field work of a most difficult and arduous nature.
D'Anville had had no such training. He was
simply a Parisian student all his life. But Rennell
looked up to him as the father of scientific geography, disagreed in his conclusions with diffidence,
and always spoke of him with the greatest respect.

He said : "If M. D'Anville is not always right, he is for the most part nearer to being so than others." *

The work of Gosselin on the geography of the ancients was also of great service to Rennell in completing his own studies, to fit him to be a worthy successor of D'Anville, and to make him competent to identify himself with the same scientific methods.

There was another French worthy to whom Rennell was greatly indebted, and one for whom he probably had a warmer feeling of sympathy than for D'Anville, though not greater admiration and respect ; for M. D'Apres was a sailor and a surveyor. Son of a captain in the French East India Company's service, Jean Baptiste Nicolas Denis D'Apres de Mannevillette—to give him all his names—was born at Havre in 1707. When he was only twelve years of age he made his first voyage to India with his father, and was afterwards sent to study mathematics and geometry at Paris. In 1726 he again embarked as an officer in one of the French company's ships, and, like Rennell, was devoted to marine surveying and to topography from first going to sea. He corrected numerous positions in the Eastern seas, and was the first Frenchman to use Hadley's quadrant. From 1735 to 1742 he was constructing corrected charts, and in 1745 he published his "Neptune Orientale," with sailing directions. M. D'Apres was in the habit of taking lunars for his longitudes. In 1749 he was captain of the *Glorieux*, and took the Abbé La Caillé to the Cape. He was a correspondent of Alexander

---

* "Herodotus," II., p. 436.

Dalrymple, and his work in the East Indies was of great use to Rennell in the construction of his Map of Hindostan. D'Apres died childless in 1780.

When Rennell commenced his literary labours, in 1780, he at once began to study all accessible geographical works with great assiduity, until, after a few years, he had acquired a profound knowledge of a subject which, in its more active aspects, had employed and interested him from boyhood. He never ceased from these studies during his lifetime, and he eventually became the greatest of living geographers. He brought reasoning and critical powers of a high order to bear on the questions which came before him; and it was his highest merit to have appreciated and followed the methods of Delisle and of D'Anville, sometimes differing from their conclusions, generally agreeing, but always working out his geographical problems on the same scientific principles.

CHAPTER V.

WHEN Major Rennell established himself at No. 23, Suffolk Street, with his wife and three little children, he found that many old friends in India had returned before him. His old messmate, Topham, lived in Queen Street, Berkeley Square. His Dacca friends, Kelsall, Hugh Inglis, Dr. Russell, and his brother-in-law Harris, were all in England, and often in London. His wife's connection was very numerous, but he soon acquired many new friends of his own among the scientific and literary men who formed a most agreeable society in those days, and met frequently at each other's houses. Rennell was elected a Fellow of the Royal Society on March 8th, 1781, and an intimacy soon sprang up between him and the President, Sir Joseph Banks. They were nearly the same age; but while Banks was devoted chiefly to botany and natural history, and Rennell to geography, they both took a catholic interest in general science. Banks was a country gentleman of Lincolnshire, with a large fortune, and he was an extensive traveller. He had been to Newfoundland with Captain Phipps, to Iceland with Dr. Solander, and he accompanied Captain Cook on his first voyage round the world. He became President of the Royal Society in 1778, and was created a baronet in 1781. He was a near neighbour of Major Rennell, his house in Soho Square being the resort of all the leading literary and scientific men of the day.

Rennell joined the Royal Society at a time of great trouble. When Sir Joseph became President, he found that the secretaries were in the habit of assuming presidential functions, and that there were other abuses which he determined to correct. There were many stormy meetings, which ended in the resignation of Dr. Hutton, the Foreign Secretary. Dr. Horsley (afterwards Bishop of St. Davids) was the leader of the malcontents, and increased the acrimonious character of the dispute by speeches of extreme bitterness. Dr. Horsley and other dissentient Fellows at last left the Society, when harmony was restored, and the ascendancy of Sir Joseph was never again questioned. The dispute appears to have resolved itself into a question whether the mathematical element was to maintain complete ascendancy; while Sir Joseph, although himself a naturalist, intended to keep a fair balance between the sciences. Rennell took no part in this controversy, and retained friendly relations with those engaged on both sides. Dr. Hutton was a very eminent mathematician, and Bishop Horsley was the editor of the works of Sir Isaac Newton, and after holding the see of St. Davids, was Bishop of Rochester and Dean of Westminster until 1802. He was a man of great intellectual force, but irritable and dictatorial.

It was through Dr. Horsley that Rennell became acquainted with Dr. Vincent, the Head Master of Westminster, and afterwards successor to Horsley

as Dean.   They had many of their tastes in common;
and the author of the "Voyage of Nearchus" and
the "Periplus of the Erythræan Sea," welcomed an
intimacy with a sailor who had made surveys on
the Indian Ocean, and in whose pursuits and ideas
the accomplished Dean had so much in which he
could sympathise.   Among his other intimate friends,
who were also neighbours, were Sir Everard Home
and Dr. John Hunter the great physicians, Alexander Dalrymple the hydrographer, and William
Marsden the historian of Sumatra and editor of
"Marco Polo"; and Lord Spencer and Lord Morn-
ington (afterwards Marquis Wellesley) in later years.
Another very intimate friend was Dr. John Gillies,
the Historiographer of Scotland, who, like Sir
Joseph Banks, was a neighbour, and nearly of the
same age.   He had a house in Portman Square;
and although he did not become historiographer
until after the death of Dr. Robertson, in 1793, he
was always devoted to literary pursuits, especially
to the study of classical history, and had many
tastes in common with Major Rennell.   He was a
good scholar, and, like Rennell, was an indefatigable
reader, but was without the spark of genius and
the critical insight which distinguished his friend.
His "History of Greece," which was much in vogue
in the end of the last century, first appeared in
1786, and was followed by a history of the world
from Alexander to Augustus, and a view of the
reign of Frederick II. of Prussia.   He also translated Aristotle's "Ethics" and "Politics," with notes.
Dr. Gillies survived his friend Rennell, and died, at
the age of ninety, in 1836.   Besides Dr. Gillies,
many . distinguished . men. of .letters. and of science

who resided in London, and numerous others, soon began to seek the acquaintance of the great geographer ; while the house in Suffolk Street was a place of meeting for eminent Indians and for travellers in all parts of the world.

With these surroundings Major Rennell commenced his first great literary work, and the construction of the first approximately correct map of India. The map was accompanied by a memoir, containing a full account of the plan on which it was executed, and of his authorities. The first edition, dedicated to Sir Joseph Banks, was published in 1783, the map consisting of two large sheets on a scale of one inch to a degree, with 146 pages of letterpress. The second edition of 1788 was considerably enlarged ; and the memoir for the third edition, which appeared in 1793, consisted of 604 pages, the scale of the map, in four sheets, being extended to an inch and a half to a degree.

It was an important undertaking. Rennell's plan was, following the principles of D'Anville, to collect all the information that was accessible to him, to discuss all the details with the greatest care, bringing all the acumen of a thoroughly logical mind to bear on the decision of each doubtful point, and to give reasons for his decisions, and a full account of his authorities in the memoir. In the last edition he included the region between the head waters of the Indus and the Caspian. The work at once secured its just place in the world's estimation, and it was many years before it was superseded by the more accurate trigonometrical survey. The map was of special value at the time it was published, supplying a want that was much

felt. Rennell himself wrote—" Now that we are
engaged either in wars, alliances, or negotiations,
with all the principal powers of India, and have
displayed the British standard from one end of it
to the other, a map of Hindostan, such as will explain the local circumstances of our political con-
nections and the marches of our armies, cannot but
be highly interesting to every person whose imagination has been struck by the splendour of our
victories, or whose attention is roused by the present
critical state of our affairs in that quarter of the
globe."

The first, and perhaps the most congenial, part
of Rennell's task was the more correct delineation
of the coast-line of India; and here he found much
valuable assistance and many helpful coadjutors.
First and foremost was his never-failing friend, the
Hydrographer to the East India Company. Alexander Dalrymple was the seventh child of Sir James
Dalrymple, of New Hailes, and was born in 1737,
being thus five years older than Rennell. He went
out as a writer to Madras in 1752, and in 1759 he
was sent on a voyage to the eastern islands. Surely
there must be something very attractive to intelligent youths in hydrography and marine surveying.
Dalrymple also became an enthusiast for this branch
of work; and when he went home on leave in 1765
he published several of the charts and plans he
had collected. In 1776 he went back to Madras as
a Member of Council, and finally returned home in
1777. There was some idea of giving Dalrymple the
command of the expedition which was more wisely
entrusted to the illustrious Cook, but in the same
year the Madras civilian received the appointment

of the East India Company's Hydrographer. In 1795 he also became the first Hydrographer of the Admiralty: a post which he held until within a few months of his death, in 1808. In 1770 Dalrymple published his "Historical Collection of Voyages in the Pacific," containing translations from the Spanish of great value. He made a prodigious collection of charts and plans of harbours and coasts in India and other parts of the East. In three volumes belonging to the library of the Royal Geographical Society there are fifty-eight charts, 740 plans, and fifty-seven views, all engraved by Dalrymple, besides fifty nautical memoirs; and these are but a fraction of his collections: there are also two folio volumes of his charts at the India Office; so that he was the great repository of hydrographic knowledge, and his information and cordial assistance were ever at the service of his friend and colleague, Rennell.

There were also three or four marine surveyors who had been bred as sailors, and were ready with their help. Of these, John Ritchie was employed from 1770 to 1785 in surveying the coast of the Bay of Bengal and the outlets of the Ganges. Many of his charts were engraved by Dalrymple, and there is a manuscript volume of his remarks at the India Office. Another surveyor of those days was, like D'Apres, Rennell, and Dalrymple, an enthusiastic hydrographer from his boyhood. This was Joseph Huddart, the son of a shoemaker at Allenby, in Cumberland, and almost an exact contemporary of Rennell, having been born in 1741. He began his sea life in the herring fishery, and from the first he showed a very remarkable natural

talent for surveying and chart-drawing. Gradually
working his way up in the world, Huddart was at
last able to build his own ship and make voyages
to the East. It was then that the East India
Company secured his services as a marine surveyor
and draughtsman from 1774 to 1790. He fixed the
longitude of Bombay by eclipses of Jupiter's satel-
lites, and his observations for latitude and longitude
along the coast of Malabar were of great use to
Major Rennell in correcting that coast-line. Captain
Huddart eventually became a director of the East
India Company, and died in 1816. Captain Michael
Topping made a chart of the Bay of Bengal in
1788; in 1790 he was employed to make a survey
of the Godavari River: a service which he performed
most creditably; and in 1792 he was taking observations for determining the course of the currents
in the Bay of Bengal. He died of fever at Masuli-
patam in 1796, where there is a monument to his
memory, with a Latin inscription. Another surveyor,
whose romantic adventures are related in my
" Memoir of the Indian Surveys " was Captain
McCluer, who did some good work in the direction of
the Persian Gulf in 1787, and on the west coast of
India. The coast-line of Tinnevelly was adopted
from the survey of Colonel Call, and Colonel Pearse
furnished correct positions from Balasore to Madras.
For improved delineations of the islands Rennell
was indebted to Captain Ritchie for the Andamans,
and to the " Neptune Orientale " of D'Apres for the
Maldives and Laccadives. With the help of these
various authorities, Rennell was enabled to draw the
coast-line of India with an accuracy such as had
never been approached before.

For the divisions of Hindostan Major Rennell adopted the Subahs, or provinces during the time of the Emperor Akbar, as defined in the Ayin Akbari. This valuable work, since translated by Blockmann, was not then accessible in a translated form, although such an undertaking had been announced by Mr. Gladwin, under the auspices of Warren Hastings. Major Rennell, therefore, obtained aid from Mr. Boughton Rouse, who translated the passages he required from the Ayin Akbari. The decision to divide the map of Hindostan primarily into the Subahs of Akbar was certainly judicious. Akbar's reign was exactly contemporaneous with that of our great Queen, overlapping it for a few years at the beginning and at the end. It was the period of the greatest prosperity and highest civilisation for Muhammadan India; and the divisions for the administrative purposes so well described by Akbar's famous minister, Abul Fazl, are of the greatest historical interest. But there were still stronger reasons for showing these divisions on the new map. The Subahdars, in some instances, formed independent kingdoms when the Mogul Empire fell to pieces, while the Subahs were long recognised as political divisions: a knowledge of which was useful even in Rennell's time. Later political divisions, such as the territories of the Mahratta chiefs and of Hyder Ali, were also shown on the map.

D'Anville's map of India had appeared in 1757, and it is very interesting to examine the points on which D'Anville and Rennell are not in agreement. Both applied the same methods of criticism to the maps and the information that was before them, and their differences arose from Rennell possessing more recent materials, while he no doubt had a great advantage, because his training was not confined to the study, but was mainly based on operations in the field.

One point of disagreement was respecting the position of Palibothra, the ancient capital of the Buddhist kingdom in the Ganges Valley. D'Anville placed it at the junction of the two rivers Jumna and Ganges, near the present city of Allahabad. Rennell, with a better knowledge of the ground, and by comparing actual distances with those he found in Pliny, saw that Pataliputra (the Palibothra of the Greeks) could not have been at the junction of the Ganges and Jumna, but that it must have been some distance lower down the Ganges. He placed it at Canouj, near Patna.

D'Anville's ground for his conclusion was that Palibothra was said to be at the junction of a large river with the Ganges, and because, according to Pliny, the Jomanes (Jumna) traversed the country of Palibothra. Rennell objected to this argument that Pliny, in another passage, assigned for the place of Palibothra a spot 425 Roman miles below the confluence of the Jumna when giving the distances of places between the Indus and the mouth of the Ganges. Using these distances, Rennell placed Palibothra at Canouj, near the confluence of the Calini, or Kalli-nadi, and the Ganges. He did not doubt that some very large city stood in the position which Pliny assigned to Palibothra, and inclined to think that this city was on the site of Patna, though he supposes the true Palibothra to be Canouj, which is below Patna and near the town of Barr, on the right bank of the Ganges.

Now, the result of the most recent researches
proves that D'Anville was quite wrong, and that
Rennell had the correct clue in his hand and was
very nearly right. Mr. Waddell has, within the last
few years, supplied conclusive evidence that the site
of Pataliputra, the famous capital of the great
Emperor Asoka, was at Patna. Rennell was the
first to point out that the Son River, or one of its
main branches, formerly joined the Ganges immediately below the modern city of Patna. Mr. Waddell
has shown that Pataliputra was on a site on the
neck of land between the old course of the Son
and the Ganges. The account of Megasthenes, the
ambassador of Seleucus Nicator, as preserved in
Strabo, is that Palibothra, which he visited in 312
B.C., was on the south bank of the Ganges, at the
confluence of another large river called the Errano-
boas, which is identified with the Son by Sir
William Jones. About five centuries from Asoka's
time Palibothra ceased to be the capital of the
Magadhan kings; it was a seat of learning at the
time of the visit of Fa Hian, the first of the Chinese
pilgrims, but when the second pilgrim, Huien
Tsiang, was there, in 635 A.D., it had been long
deserted, and was a mass of crumbling ruins. In
the time of Akbar the town of Patna became the
largest in the province, and the Buddhist relics of
Pataliputra were gradually buried in the *débris* of
many centuries. Mr. Waddell published the interesting narrative of the discovery of the ruins of the
principal edifices of Palibothra at Calcutta in 1892.

Another point of disagreement between D'Anville
and Rennell had reference to the course of the
Tibetan River Sanpu. It is shown on the map of

Tibet drawn from surveys made under the auspices of the Jesuits, and afterwards published by Du Halde. Bogle, who was sent on a mission to the Teshu Lama in 1774 by Warren Hastings, crossed the Sanpu twice. He cleared up a few points, but he was not a surveyor, took no bearings, and observed for no latitudes. The mighty chains of the Himalayas, separating the valleys of the Sanpu and the Ganges, were called the Tartarian Mountains when Rennell first saw them in the course of his survey, and in the memoir he speaks of them as a continuation of the Emodus and Paropamisus Mountains of the ancients, called by the Tibetans "Rimola." On the map, however, the name Himalaya appears, and the mountains are spoken of as equal in height to any mountains of the old hemisphere. The Sanpu was known from the map in Du Halde, and from the testimony of Bogle and Turner, to flow eastward, to form an affluent of some great river emptying itself into the sea. D'Anville held that the Sanpu was a tributary of the Irawadi. But Rennell, after considering every point bearing on this difficult question, and with truer geographical insight, came to the conclusion that the Sanpu pierced the Himalayas, and, as the Dihong, flowed into the Brahmaputra. This was long a vexed question. Mr. Gordon, who was employed on the survey of the Irawadi in later times, adopted the view of D'Anville; but the great weight of authority is on the other side, and now there is practically no doubt that Rennell was right, and that the Sanpu is a tributary of the Brahmaputra.

In these two instances of the site of Palibothra

and the course of the Sanpu, the critical judgment
of Rennell was superior to that of D'Anville. Both
were the most eminent geographers of their respective times, and both used similar methods in
considering questions relating to geographical subjects. The difference between these two eminent
men was that while one was a student from his
boyhood, the other had been a surveyor on active
service by sea and land for nearly a quarter of a
century before he commenced his literary career
No doubt this experience in the field had an important influence in guiding the student to accurate
judgments in after life, and in giving him an insight
which was not possessed by the life-long worker within
the four walls of his study. The conclusion is that
to form a perfect geographer long experience in the
field as an explorer or a surveyor must be combined
with the close and deep study of a man of letters.
Major Rennell possessed both these essential qualifications; and while making himself acquainted with
the authorities of former ages, he himself had the
power to decide questions on his own authority.
He, too, had traversed wild and uninhabited forests,
had followed the courses of unknown rivers, and
had measured vast regions before unmapped, and
in the course of this work he had acquired an insight, and indeed an instinct, which no mere scholar
could ever hope to possess.

In the delineation of the provinces of Hindostan,
Rennell had his own careful and elaborate survey
for Bengal, Bahar, and Orissa. He used the Persian
map of Sir R. Baker for the names—which were
translated for him by Major Davy—in the Indus
Basin, Rajputana, and the Punjab, adding other

names, obtained from the Ayin Akbari, Ferishta, and Sherrif-u-din. He also studied the routes of the several invaders of India, and adopted the details in Kashmir, and the route from Ajmir to Jesalmir from D'Anville's map.

In Central India and the vast region from the Ganges to the Kistna, Rennell trusted to the marches of M. Bussy, to information from Colonel Camac respecting the roads between Bengal and the Deccan, to details of the march of General Goddard from Calpi to Surat, and to the routes of Captain Price. He referred to the earlier journeys of Bernier and Tavernier, and from this part of his work he also derived occasional assistance from D'Anville's map. The groundwork of his map for the country south of the Kistna was derived from the map in Orme's work, part of which was based on the surveys of Colonel Call. Rennell also had the use of Mr. Montresor's manuscript map at the India House, and derived further details from the marches of Colonel Fullerton in Coimbatore, and of other officers during the wars with Hyder Ali and Tipu.

In the last edition of his map, Major Rennell was enabled to add valuable geographical details respecting the country between India and the Caspian Sea from the information brought home by Mr. Forster, which was communicated to him by Lord Mulgrave. Forster afterwards published his travels in two volumes, and they formed the best authority for that region until the appearance of the works of Sir John Malcolm and the memoir of Kinneir.

There were two other friends whose judicious

advice and ever-ready assistance were of great help
to Major Rennell in the execution of his famous
map of Hindostan. Mr. Wilkins went out to Bengal
in the Civil Service in 1770, and devoted himself
to the study of the Sanscrit language. He translated a version of the Bhagavat Gita, to which
Warren Hastings prefixed a dissertation, the whole
being published by the Court of Directors in 1783.
Returning to England in 1786, Wilkins made a
translation of Sakontala in 1795, and published a
Sanskrit grammar in 1808. He became Librarian
at the India House in 1800, and when in extreme
old age he was knighted by William IV. Sir
Charles Wilkins died at his house in Baker Street,
aged eighty-six, in May, 1836.

Rennell's other helpful friend, while he was engaged upon his map of Hindostan, was the son-in-
law of Sir Charles Wilkins. William Marsden was
an Irishman from Wicklow, and was educated at
Trinity College, Dublin. He obtained a writership
at Bencoolen, in Sumatra, and sailed from Gravesend
in 1770. After eight years of service he became
Secretary to Government, returning to England in
1779, where he established an East India agency,
with his brother John, in Gower Street. Marsden
was elected a Fellow of the Royal Society in 1783,
became a Vice-President, and often presided at the
meetings. He was also Second Secretary to the
Admiralty from 1795 to 1807, retiring with a pension of £1,500 a year, which he afterwards resigned.
Author of the "History of Sumatra," of a Malay
grammar and dictionary, and of the *Numismata
Orientalia*, and editor of the travels of Marco Polo,
Marsden held a good position as a leading man of

letters. He was popular, and possessed conversational powers of a high order, and was generally to be met at Sir Joseph Banks's philosophical breakfasts in Soho Square. Marsden lived to be a founder of the Raleigh Club, the precursor of the Royal Geographical Society, and survived until 1838.

Wilkins and Marsden were constant in their helpful advice; while Mr. Beasley and Sir Hugh Inglis among the directors, Mr. Sulivan of Madras, Mr. Callander of Bombay, and many others, were ever ready with their local knowledge. In 1781 Major Rennell had communicated a paper giving an account of the rivers Ganges and Brahmaputra to the Royal Society, which afterwards formed an appendix to his memoir on the map of Hindostan. He also wrote a paper for the *Philosophical Transactions*, " on the rate of travelling by camels," and one on " the marches of the British army in Central India in 1790-91."

For his work respecting India, Major Rennell was adjudged the Copley Medal of the Royal Society in 1791; and the President, Sir Joseph Banks, in his address, thus gave his estimate of the value of that work :—

"I should rejoice could I say that Britons, fond as they are of being considered by surrounding nations as taking the lead in scientific improvements, could boast a general map of their island as well executed as Major Rennell's delineation of Bengal and Bahar: a tract of country considerably larger in extent than the whole of Great Britain and Ireland ; but it would be injustice to the Major's industry were I not here to state that the districts he has perambulated and planned exceed, probably, in extent the whole tract of surveyed country to be found in the maps of the European kingdoms put together,

while the accuracy of his particular surveys stands yet un-
rivalled by the most laborious performance of the best county
maps this nation has hitherto been able to produce." *

The map of Hindostan, with the memoir, at
once established Rennell's fame as a geographer and
a man of letters. The "Copley Medal" was a
greater honour than anything the Government
could confer, and he was treated by the scientific
and literary men with whom he chiefly lived with
a diffidence and respect which must have shown
him how highly his meritorious services were appreciated by those whose opinions were most
valuable. This is also observable in the conduct of
the Court of Directors, and from this time Major
Rennell, sometimes in conjunction with Dalrymple,
became their adviser in all questions relating to
geography and marine surveying.

In September, 1791, the Major was requested by
the Court to examine and report upon a map of
Hindostan in thirteen parts, drawn in India by
Colonel Call, who had been a successor to Rennell
as Surveyor-General of Bengal, and was then dead.
Rennell was unable to recommend that it should
be engraved, because it was not by any means up
to date. It did not contain the various route
surveys of Captain Reynolds, those of Captain
Beetson, nor the marine surveys of McCluer and
Topping. These materials had reached the India
House, and some of them were included on published maps, but none of them could be found on that
of Colonel Call. Valuable new materials were also

---

* Given by Baron Walchenaer in his *éloge*.

G

immediately expected from India; so that Colonel
Call's map would be obsolete before the engraving
was completed. Major Rennell's duty to the Court
obliged him further to point out that there was no
history of the construction of the map, nor statement of the comparative value of the different
materials that were used. Outside actual surveys,
" much the greatest proportion of the materials
must rest on the foundation of judgment or
opinion. We have no means of discriminating their
value, or of knowing how far the positions that lie
outside the surveys depend on the coincidences of
lines of distance and bearings, or whether regulated
by observations for latitude or otherwise. In this
respect the matter can only be compared to a long
and intricate account, without vouchers or explanation. This defect may, however, be remedied by
sending for the original papers themselves."
Colonel Call's map was never engraved, and I
believe that it is now lost.

Another question of importance was submitted
by the Court of Directors for Major Rennell's
opinion, having reference to some proposals made
by General Roy for geodetic experiments to be
made in the East Indies, with an account of the
mode proposed to be followed in the trigonometrical
operation for determining the relative situations of
the Royal Observatories at Greenwich and Paris,
and observations on the magnitude and figure of
the earth. The Court professed itself extremely
desirous of affording every facility towards the
success of General Roy's plans and experiments,
and Major Rennell reported strongly in favour of
complying with his wishes. General Roy commenced

the trigonometrical survey of Great Britain and Ireland a few years afterwards. The Court of Directors consulted Major Rennell on other points of a like nature from time to time, and he was always their adviser, in conjunction with Dalrymple, on matters relating to hydrography.

The survey of Bengal and Bahar, the "Bengal Atlas," and the map of Hindostan, with the memoir, represent services such as have seldom been conferred on a country by one man. They received no direct official recognition whatever; but the author was rewarded in a way which, to a man like Rennell, was far more acceptable. It consisted in the cordial appreciation of his countrymen at home and in India, in the respect and esteem which his labours at once secured him, in the acknowledgment made by the Court of Directors in constituting him their geographical adviser, and in the honours conferred upon him by scientific bodies at home and abroad.

CHAPTER VI.

AFTER the completion of his geographical work for India—which includes the survey of Bengal and Bahar, his "Bengal Atlas," his map of Hindostan, and his memoir on the geography of India—Major Rennell turned his attention to Western Asia: that is, to the portion of that continent between India and the Mediterranean. He had conceived a very comprehensive scheme for a great work on the comparative geography of Western Asia; and his geography of Herodotus, only forming a part of the whole project, occupied him during many years.

In those years he continued to live in the house at No. 23, Suffolk Street, enjoying much domestic happiness with his wife and three children. Mrs. Rennell was the main source of this happiness. She was beloved by her relations, and numerous acts of affection and kindness are recorded of her. In London she was a weekly visitor to the poor at Middlesex Hospital. But latterly the winters were usually passed at Brighton, and there were occasional tours to Wales, when the Menai Straits were crossed, and to Scotland, always with the children. During one tour in Scotland Major Rennell broke his wrist, owing to his horse falling with him. Miss Rennell, who was born at St.

Helena in October, 1777, grew up to be a very
handsome girl, and a source of pride and happiness to her father. The two boys, Thomas and
William, were very promising scholars. They were
educated at Dr. Horne's school at Chiswick. The
eldest went thence to Trinity College, Cambridge,
and was sixth Wrangler in 1801. William, the
youngest, entered the Civil Service of the East India
Company, and was magistrate at Dacca, where his
father had so long resided. In mentioning those to
whom he was indebted for assistance, in the preface
to his geography of Herodotus, Major Rennell
had great pride in being able to include his two
young sons.

The years during which Rennell was engaged
on his great work relating to Western Asia were
those which saw the outbreak and mad career of
the French Revolution. The illustrious geographer
was always an advanced Liberal in politics, but he
loved his country's greatness above all things, and
imbibed a strong feeling of horror for the excesses
of the French Revolution. He consequently attached himself to those Liberals, such as Lord
Spencer and Sir W. Windham, who joined Burke
in supporting the government of Pitt in a war
policy. He was the author of a pamphlet, in 1794,
entitled "War with France the only Security of
Britain," and he expressed his opinions very clearly
in the dedication of his "Herodotus" to his friend
Lord Spencer. "Some parts of the work," he
wrote, "may recall ideas respecting the history and
policy of those nations of antiquity whose learning and arts we are ambitious of imitating, and
whose liberty is a perpetual theme of praise even

amongst us who have employed ages in perfecting
a practical system of our own, which, although
subject to decay, like all other human institutions,
promises to be of much longer duration than any
other on record." He then refers to the liberal
earl's coalition with Pitt, and to his organisation
of victory, which led to St. Vincent and the Nile.
"To preserve this wonderful fabric entire in all its
parts, your Lordship joined your counsels and
exertions at a momentous crisis. History will relate the acts of your department; that from the
Ganges to the Nile, and from the Nile to the
shores of the sister island, the desperate projects of
the inveterate enemy of mankind against the safety
and interests of this empire were totally frustrated."

Similar thoughts could not fail to arise in the
mind of this patriotic Englishman and friend of
Lord Spencer, when the course of his studies of
Herodotus brought him to the Canopic mouth of
the Nile. "The classic importance of Canopus," he
wrote, "is very great, considered either as a place
visited by the heroes of the Trojan war, as the
reputed burial-place of the pilot of Menelaus, or
in respect of the rank it held amongst the cities of
Egypt. But as some ancient places have been so
fortunate as to renew their classic importance in
modern times, as if to ensure the certainty of a
longer term of celebrity, so Canopus, under the
modern name of Aboukir, has received a new and
perhaps a more lasting impression of 'the stamp of
fate,' by its overlooking, like Salamis, the scene of
a naval battle which, like that of Salamis, may
lead to a decision of the fate of Europe. This
most brilliant victory, achieved solely by Britons,

Europe felt as her own. To this spot the genius of Britain conducted her favourite Nelson, who at one blow destroyed the fleet of the enemy, and cut off the veteran army of the French from their country.

"But what secluded shore of the ocean has not, in its turn, reverberated the British thunder? During the present struggle, what walls have resisted save the wooden walls of Britain? Nor shall History, although she delights more to record a brilliant victory than the councils that produced it, fail to hand down to posterity the name and character of the Naval Minister * who so successfully directed the great engine of British power! Devoted to her service, his country shall claim him as her own to the latest times; whilst France shall recognise in the descendant of Marlborough the hereditary foe to her schemes of ambition and aggrandisement."

In another place Rennell's political reflections, suggested by the Grecian resistance to Persian invasion, breathe the same spirit. "In the history of the Persian invasion and its termination, so glorious to Greece, Herodotus has given a lesson to all free States that either do exist, or may hereafter exist in the world, to dispute their independence, let the number of their enemies be what it may. He has shown that the Greeks, although a large proportion of their country was in the hand of the enemy, were still formidable, and in the end prevailed over a foe that outnumbered them more than three to one in the decisive battle of Platæa; notwithstanding

* Lord Spencer.

there were included in that vast majority as many of
their renegade countrymen as amounted to nearly
half their own numbers.   The event of the contest
depended chiefly on the obstinate determination of
the Greeks not to submit: a resolution which, when
accompanied by wisdom and discipline, must ever
prevail.

"The Dutch acted like free men when they determined to defend their last ditch against Louis XIV.,
and, in the last resort, to embark for their foreign
settlements, as the Phocœans aforetimes did for
Corsica.   The Anglo-Americans have just displayed
the same noble sentiments (worthy of their ancestors)
in treating with equal contempt the proffered friendship and the threatened enmity of France: France,
whether monarchical or republican, the common
enemy of the peace and independence of nations."

Referring to the great English colonial empire, he
wrote:—" In the prospect of future times there is a
subject for pride in the breasts of Englishmen, which
is that so vast a portion of the globe will be peopled
by their descendants.   I allude, of course, to America
and New Holland.   The latter alone appears to have
room enough for as many inhabitants as Europe at
present contains.   That is, at least, beyond the power
of the French Directory to prevent, for the progress of
population is too rapid to be opposed by human
means, and will soon outgrow, in America, that of
France, with all her conquests and fraternisations.
The colony of New South Wales, too, will probably
be able to take care of itself.   The rising generation
there is said to be very numerous, and it is pretty
obvious that on the care of their religion and morals the
character of the future nation will depend.   It ought,

perhaps, to afford a triumph to literary men to reflect
that the English language had received its highest
degree of improvement before the epoch of our great
colonisations. He, therefore, who writes in English,
and whose works descend to posterity, will probably
have the greatest number of readers. This was,
perhaps, the case heretofore of him who wrote in
French."

Major Rennell's political opinions are easy to be
known from these extracts and from passages in his
letters. He was a Liberal who sympathised warmly
with the working classes, and who loved alike the
independence and freedom of his country and the
individual liberty enjoyed by his countrymen. He
was zealous for the glory of England and for her
rightful position among the nations, while, in common with all the best and most patriotic men of his
generation, his horror of the principles of the French
Revolution, which seemed to have led to such hideous
excesses, was slightly exaggerated. He saw things
with the vivid distinctness of proximity in time and
space, while we are able to view them calmly and at a
distance, and to weigh the relative strength of various
causes which brought about such results. With pro-
phetic vision Rennell seems to have made a forecast
of the future greatness of the United States and of
the Australian colonies, and to have anticipated the
marvellous increase in population and material prosperity that has since taken place, and the eventual
predominance of the English language.

His study of comparative geography was far from
unfitting him for forming sound judgments on
political questions. In such times no one could remain indifferent; and Rennell, who through life had

taken such a keen and intelligent interest in passing
events, least of any one.   Indeed, his occupations, his
habit of carefully weighing evidence and of deciding
between conflicting authorities, his research and extensive general knowledge, gave his opinions more
than usual weight.   At the same time, deep and
intense as was his interest in all that affected the
welfare and glory of his country, he did not swerve
from the useful occupations to which he had devoted
his life.   The correction of the ancient and modern
geography of Western Asia was indeed a serious
undertaking.   This region was alike the great theatre
of sacred and profane history in ancient times, and of
modern commerce and communication.   Rennell's
first object was to adapt his system to the use of
statesmen and travellers; and his secondary aim was
to apply it to the illustration of such parts of ancient
military history as were deficient from the want of
necessary geographical aids, which have, in some
degree, been supplied in modern times.

The first part of the undertaking was his work on
the geography of Herodotus, forming a complete work
in itself.   The other divisions were intended to be a
work on ancient geography as it was improved by
Grecian conquests and establishments, and an atlas of
ancient geography, each also forming complete works
in themselves.   As regards the first division, he had
completed his task, so far as his stock of materials
admitted, by the last year of the last century.   The
first edition of his "Herodotus" was published in 1800.
The second revised edition was brought out by his
daughter, Lady Rodd, in 1830.   Rennell was not
acquainted with the Greek language, and could only
acquire a knowledge of the text of Herodotus by

means of a translation. He wrote :—"The magnitude
of this defect will perhaps be differently estimated by
different persons. It may doubtless be said, with
truth, that no ordinary reader of Greek is likely to be
so perfect a master of Herodotus through a perusal of
the original work as by translations made by professed
scholars who have devoted a great portion of their
time to the study of it. On the other hand, it must
be allowed that such scholars, if also skilled in the
science of geography, would be by far the fittest
persons to undertake a task of this kind. Such
an one, however, has not yet undertaken it, and
therefore the author flatters himself that, in the
existing state of things, his work may be allowed
to pass until so desirable a coincidence may take
place."

The translation used by Major Rennell was that
by Mr. Beloe: the only one in existence at that time.
Unfortunately, Beloe's version was very inaccurate ;
and this gives an opportunity for the critic in the
"Penny Cyclopædia"* to bestow the highest praise on
Rennell. He wrote : " Though obliged to trust to the
very inaccurate version of Beloe, Major Rennell succeeded in producing a commentary on a classical
author which is not surpassed by the labour of any
scholar. The blunderings of Beloe, and his occasional
complete perversion of the original, did not mislead
the geographer, who could detect the author's meaning under the disguise of the translation." Another
writer observed that " Major Rennell only misunder-
stood Herodotus when Beloe deceived him, and
frequently, as in the case of the Tower of Belus, he

* Quoted by Sir Henry Yule.

found out the meaning of the author in spite of the translator " *

As in his memoir on the map of Hindostan, so in his work on Herodotus, Major Rennell received constant help and critical advice from such friends as Dalrymple, Wilkins, Marsden, Gillies, and many others. Rennell had selected Herodotus as the best exponent of geography in the point of view of a starting point, because, as his writings furnish the earliest record of history, so they also supply the earliest known system of geography, as far as it goes. The geographical notices of the father of history are scattered throughout his work, always being placed where they may best serve to elucidate the parts of history to which they respectively belong, and not with a view to an abstract system of geography. Rennell's plan was to collect all the scattered geographical notices in Herodotus into one point of view, in order to make them bear on and illustrate each other systematically. Herodotus had the advantage of having visited the countries he particularly describes. His notion was that the known world, consisting of Europe, Asia, and Africa, was surrounded by the ocean, except towards the east, where he believed there was a vast and unexplored desert of unknown extent. On some points his knowledge was superior to that of writers who lived much later. For instance, Herodotus knew that the Caspian was an inland sea; while, from the time of Alexander to that of Ptolemy the geographer, the belief was that the Caspian was a gulf of the northern ocean.

After his introductory observations, Rennell

* *Quarterly Journal of Education*, I., p. 329.

devoted a chapter to the discussion of the length of the itinerary stadium of the Greeks : a most important question, upon the correct determination of which the whole fabric of arguments and deductions mainly depended. It was a subject on which D'Anville had already written a learned treatise. Herodotus described the stadium as a measure of 600 Grecian feet, or about 600 to a degree. This was the Olympic stadium ; but the itinerary stadium used in describing routes appears to have been shorter than the Olympic stadium. That of Xenophon was 750 to a degree, of Strabo 700 to a degree; and Herodotus evidently used more than one standard for the itinerary stadium : a longer one in Greece and Persia than in Egypt and Euxine Scythia. In carefully comparing standards of the chief authorities of antiquity, Rennell found that the mean of all was 718 stadia to a degree : equal to 500 English feet. He noticed that the double step pace was equal to five feet, and thought it probable that the original stadium was a hundred of these paces, or 500 feet. Herodotus and others probably always intended the same stadium, but they may have given occasion to different results by their having reported the numbers on the judgment of different persons.*

* Colonel Martin Leake, in a paper read before the Geographical Society in 1838, fully discussed the stadium as a linear measure. He showed that the Attic foot, taken from the stylobate of the Parthenon, was equal to 12·1375 English inches, and that the length of the stadium was, therefore, 606·875 English feet, or 625 Roman feet : equal to a furlong, or one-eighth of a Roman mile. According to Herodotus, the Egyptian cubit was equal to the Samian stade. The Egyptian cubic is 20,$\frac{6}{10}$ English inches, so that the Samian foot would be 13$\frac{3}{4}$ English inches, and the Samian stade equal to 687 English feet. There was also a Pergamenian stade of 697 English feet ; but these are the only varieties of stades for which any support can be found in

After showing how very little Herodotus knew of
Western Europe, Rennell enters upon a full discussion
of the geography of Euxine Scythia from the Danube
to the Don, and northwards to the Volga, and of the
march of Darius Hystaspes. Speaking of the ancient
province of Hylœa, adjacent to the Tauric Chersonese,
he alludes to the changes in the course of the Borysthenes since the time of Herodotus, as well as the
coast between the mouth of the Borysthenes and the
Crimea, referring frequently to such modern authorities as Baron Tott and Pallas. This is one among
many striking examples of the value of an ancient
authority, such as Herodotus, in the study of physical
geography, by enabling us to compare the condition
of a particular region at periods two thousand years
apart. Another example may be mentioned at the
Pass of Thermopylæ, in comparing the descriptions
of Herodotus and Colonel Leake. It is true that
Herodotus, as a rule, does not give particular geographical descriptions of countries which are supposed
to be well known to his readers, but he does so sometimes; and Major Rennell selects his description of
Thessaly as one of the most pointed, clear, and concise
imaginable.

In describing the country of the Budini and Geloni,
on the Upper Tanais, he identifies it with that of
Woronez, on the Don, and takes occasion to refer to
the account given by the traveller Le Brun of a fleet
built by Peter the Great at Woronez. Rennell formed
a high estimate of the character of the Czar. "When,"
he wrote, " we reflect on the various personal labours

ancient history.   There would be 600 Greek stades of 600 Greek feet
to a degree.

of this truly great prince, all tending to produce either
an immediate or a remote advantage to his country—
now enforcing duty by example, now operating the
direct means of national strength or improvement;
considering also the unusual means pursued by him
to obtain the requisite degree of knowledge—we are
struck with admiration, and cannot help exclaiming,
with Addison, 'Who before him ever left a throne to
learn to sit in it with better grace ?' "

These occasional interesting deviations from the
direct course of the disquisition confer a special
interest on this work, and are useful by giving rise
to appropriate reflections and impressing the facts
connected with particular regions on the memory.

In treating of the Tauric Chersonese, Rennell has
occasion to refer to the evidence of Rubruquis ; and it
is by such references throughout the work that we
obtain an insight into the extent of his researches
while preparing to write it. The warning sent by the
King of Scythia to Darius to beware of doing any
injury to the sepulchres of his ancestors in the district
of the Geroli gives occasion for Rennell to furnish a
most interesting dissertation on these tumuli and their
contents, scattered throughout Russia, using the paper
by Mr. Tooke in the *Archæologia*, VII., p. 223, as
his principal authority. He also has a short and
interesting chapter on the bridges thrown across the
Bosphorus and the Hellespont by Darius and Xerxes,
relying upon such modern authorities as Tournefort,
Gibbon, Tott, and Pocock.

When he comes to treat of the countries beyond
Euxine Scythia, we find Rennell comparing the statements of Herodotus with such authorities as Petis de
la Croix, Abulghazi Khan, Edrisi, and Abulfeda. The

custom of the Hyperboreans of sending their offerings
to Apollo of Delos by the hands of two virgins gives
occasion for some interesting reflections, which seem
well worthy of being quoted.

On one occasion the two Hyperborean virgins
who had come with offerings to Apollo died at Delos,
and the guard that brought them never returned. To
prevent a repetition of this disaster, the Hyperboreans
henceforward sent their offerings through the agency
of intervening nations; but the Delian youth showed
their sympathy by celebrating certain rites in honour
of the virgins. Major Rennell observes: " There is
something more than ordinarily melancholy in the
fate of those who, visiting a distant country on some
specific errand, and with a view to immediate return,
perish untimely in a strange land. How often has
this happened in our own times: in particular, the fate
of Tupia* and of Lee Boo† interests us, from their
amiable dispositions and the grief of their friends who
awaited their return.

* Tupia was a chief and priest of Tahiti, who embarked with
Captain Cook on board the *Endeavour* in July, 1769, desiring to visit
England. He was very intelligent, and his services were most valu-
able as an interpreter in New Zealand; but he died at Batavia in
December, 1770.

† When the East India Company's ship *Antelope*, commanded by
Captain Wilson, was lost on one of the Pelew Islands in 1783, the
king received the shipwrecked people with great hospitality, and his
subjects assisted them to build a small vessel. Captain Wilson
offered to take the king's son, Prince Lee Boo, to England, have him
educated, and send him back. The proposal was accepted, and Lee
Boo was brought to England, where he was carried off by small-pox
in December, 1784. He was an amiable and most promising youth.
Everybody has read his story in the " Child's Own Book." The sad
news of the prince's death was brought to the Pelew Isles in 1791 by
the *Panther* (Captain McCluer).

" Whatsoever has a tendency to link mankind together in peaceful society is pleasing to liberal minds, and therefore we feel a degree of sorrow for such accidents. For whether the object of the visit be rational curiosity or harmless superstition, or both, the effect produced on the mind may be good, while the benefits that whole communities may derive from the inquiries of such travellers are, in some cases, incalculable.

" The world has seen a Pythagoras, an Herodotus, a Peter Alexionitz, and a Banks forego either the exercise of unlimited power, the blandishments of elegant society, or at least the comforts of ease and security, to brave the dangers of the deep, or those greater dangers which often arise from intercourse with man in his savage state, in quest of knowledge or of useful productions—not that kind of knowledge alone which merely administers to the pleasure of the traveller, but that which is derived from inquiries concerning what useful customs or institutions among men and what products of the earth or sea might be imported into their own countries or colonies.

" The interchange of useful vegetable products between the different countries of the earth, with a view to cultivation, is alone an object which commands the gratitude of the world; and happy the man whose fame rests on this solid foundation—a foundation that opinion cannot shake, since all feel and participate in the benefits. Systems of politics and the fame of their authors vanish, and are, in comparison to the other services, like unsubstantial clouds, that vary their form and colour with every change of position or circumstance."

There is a striking remark on this subject which

H

Dr. Swift puts in the mouth of one of his characters in " Gulliver's Travels " :—" Whoever would make two ears of corn or two blades of grass to grow upon a spot of ground where only one grew before, would deserve better of mankind and do more essential service to his country than the whole race of politicians put together."

These and other reflections which occurred to the author, as he studied the events and the anecdotes related by Herodotus, are interspersed through the work, and give it a peculiar charm ; for Major Rennell thus not only elucidates the geographical descriptions of the father of history, but allows the reader to gain an insight into his own thoughts and opinions on many subjects. Having completed his view of the geography of Eastern Europe, Rennell next gives, for the time in which it was written, an admirable general description of the physical geography of Asia, especially explaining the nature of the great central Tibetan plateau, and of the mountain buttresses which encircle it. He pays very special attention to the questions relating to the Massagetæ, Sacæ, and other Scythian tribes, comparing the statements of all the writers of antiquity on this subject with those of modern travellers, and of such oriental authors as were accessible to him. Rennell's chapters on the twenty Satrapies of the Empire of Darius Hystaspes are especially interesting. In one place there is a discussion of the method of water-supply for large armies crossing a desert; and he compares the accounts in Herodotus, especially referring to the march of Cambyses into Egypt, with the method used by Nadir Shah, as described in Mr. Gladwin's translation of " Abdul Kerreem." He compares the description of circular willow boats on the Euphrates by Herodotus, with the account of similar vessels seen by his friend Mr. Sulivan, who made a journey overland to India, and thus in many ways, and on every available occasion, he brings the corroborative or illustrative testimony of modern travellers to bear on the narrative of the father of history.

In treating of the Satrapy of Susa, Major Rennell dwells upon the fact of the large immigration which took place in the days of Persia's early greatness into this interesting province. The Prophet Daniel and a company of Jews were at Susa, and here the Eretrians of Euboea were placed by Darius, whose generals captured them, and carried them off during the first Persian invasion of Greece. This Grecian colony can be traced as inhabitants of Susiana, preserving their language, for 540 years after the first captivity. In treating of Media, Major Rennell expresses his belief that Ecbatana was on the site of the modern Hamadan, contrary to the opinions of Gibbon and Sir W. Jones. His arguments are that Susa was said to be half-way between Ecbatana and Seleucia; that Ecbatana was on the road from Nineveh to Rhages, and that it was on the road from Seleucia to Parthia. These indications point to Hamadan as the site. When speaking of the islands in the Persian Gulf, which were included in the Fourteenth Satrapy, Rennell mentions that there is much curious history belonging to these islands. They have at different times contained the commercial establishments of the Phoenicians, and also of European nations. " But what is more gratifying to the mind is that they have, in modern times, afforded asylums to the inhabitants of the maritime towns on the continent when invaded or oppressed;

and so regular has the system of taking refuge been, that some of the islands have their names from the opposite towns on the continent. In particular, the inhabitants of the continental Ormuz passed over into the island of that name (the Organa of Nearchus) on the irruption of the Tatars, in the thirteenth century. None can feel the importance of insular situations to the cause of liberty more than Englishmen, especially at this time. The Tatars had no fleets to pursue the fugitives to the islands, but the King of Persia, who possessed ships, made use of the islands as places of banishment."

The object in making these quotations is not to review Major Rennell's work, but rather to illustrate his thoughts and the workings of his mind. They may appropriately be concluded with his remarks on the intercourse between Alexander the Great and Calanus. While Alexander was in the Punjab some Brahmans were brought to him, and one of them, named Calanus, returned with him into Persia. Mr. Williams suggested to Major Rennell that the real name was Kalyanah. " In this Indian philosopher we trace, at the distance of more than twenty-one centuries, the same frame of mind and the like superstitions as in the same tribe in our own times: a contempt for death, founded on an unshaken belief in the immortality of the soul, and an unconquerable adherence to ancient customs. The friendly connection that subsisted between Alexander and this philosopher does infinite honour to both, for it proves that both possessed great minds and amiable dispositions. Alexander never appears to such advantage as during the last act of the life of Calanus. This Indian sage, finding his health decline, and

believing that his end approached, determined to
lose his life on a funeral pile to avoid the misery
of a gradual decay, to which Alexander reluctantly
consented, from an idea that some other mode of
suicide, less grateful to the feelings of Calanus, would
certainly be resorted to. Alexander even condescended
to arrange the ceremonies himself; and Arrian appears
to be much struck with the character and fortitude of
Calanus, and remarks: ' This is an example of no mean
import to those who study mankind: to show how
firm and unalterable the mind of man is when custom
or education has taken full possession of it.' " Major
Rennell refers to the story of Allavee Khan, a respectable Muhammadan physician of Delhi, who returned
to Persia with Nadir Shah, and exercised a good
influence over one of the most stubborn and blood-
thirsty tyrants the world has ever produced. But
Calanus had to deal with a conqueror who was also an
educated philosopher, and had an easier task than
Allavee Khan in managing Nadir Shah.

The two last sections, containing an examination
of the report of Aristagoras concerning the royal road
from Ionia to Susa, and a dissertation on the site and
remains of Babylon, are very important to students of
ancient history. The second volume of Rennell's
" Herodotus " treats entirely of the geography of
Africa. His paper " On the Topography of Troy,"
and his works on " Cyrus and the Retreat of the Ten
Thousand " and on " The Comparative Geography of
Western Asia," are parts of his great project, each,
like the geography of Herodotus, forming complete
works in themselves.

" The Illustrations (chiefly geographical) of the
History of the Expedition of Cyrus and the Retreat

of the Ten Thousand" was dedicated to Lord Grenville, and published in 1816. The "Anabasis" is a
military history, but it is also a book of travels, for
the Greeks went over 3,700 miles of ground, and
Xenophon described it all. By the light of the work
of contemporary travellers, and through his own
sagacity, Rennell was enabled to make numerous
important corrections. The map of Asia Minor by
D'Anville is over a degree out in latitude, which
threw out most of his positions, so that Rennell had
the task of reconstructing the map afresh. In this
difficult undertaking he received help from Von
Hammer, who sent him translations of several
Arabian geographers from Vienna. Niebuhr sent
him much information and a map of his route
through the southern part of Asia Minor. His friend
the Right Hon. J. Sulivan furnished him with most
valuable notes of a journey up the valley of the
Tigris, while Dr. Gillies was ever ready with advice
respecting Xenophon's text. The translations used
were the English version by Spelman and the French
by Larcher.

The African traveller, William G. Browne, had also
done work which was of considerable assistance in
tracing the march of Cyrus. Born in 1768, Browne
was at Oriel College, and graduated at Oxford,
devoting his future life to scientific travel. From
Egypt, he explored the oasis of Siwah, the Roman
quarries at Cosseir, and reached Darfur in 1793.
Returning to England, he published his African
travels in 1800, and almost immediately set out on
an adventurous journey through Asia Minor and
Persia to Central Asia. His routes were very carefully laid down, and sent home for Rennell's use.

They covered parts of Asia Minor in which the great geographer was most interested, and he frequently ·acknowledges the assistance derived from them. Proceeding eastward, the intrepid traveller entered Persia, where his labours came to an untimely end. Mr. Browne was murdered by brigands, in 1813, on the road between Tabriz and Tehran.

Major Rennell erroneously supposed the "royal road" to have been the same with that followed by Cyrus and described by Xenophon, so that his examination of its details is rendered valueless; but he corrected the map of D'Anville, and, with the aid of modern routes and information, he threw much light on numerous passages in the text. He took great interest in this portion of his work, and pronounced the expedition of Cyrus, taken in all its parts, to be the most splendid of all the military events that have been recorded in ancient history.

In 1814 Major Rennell published his observations on the topography of the plain of Troy, with a map. His object was to show that the then generally received system of Chevalier, brought forward in 1791, was founded on erroneous topography. The major again apologises for his want of knowledge of the Greek language, but he remarks, very truly, that it does not always happen that a critical knowledge of languages and a turn for geographical disquisitions meet in the same person. He made use of Cowper's translation of "Homer"; and received much advice and assistance from Sir Joseph Banks, Dr. Gillies, and especially from Gell, the author of the "Topography of the Plain of Troy," published in 1804.

Two volumes were published by his daughter, Lady Rodd, after his death, which were, in fact,

Rennell's workshop, displaying his critical methods
and his treatment of the materials he collected.
Their title is " A Treatise on the Comparative Geography of Western Asia, with an Atlas of Maps."
His introduction describes the authorities on whose
work he relies for the different sections of his map:
Niebuhr supplied him with a line of distances from
Hilla, on the Euphrates, to Brusa; Browne with routes
in Asia Minor; Beauchamp gave him a more correct
delineation of the south shore of the Black Sea; the
positions and itineraries of the Arabian geographers
were useful; and valuable maps were drawn by a
Hungarian renegade, named Ibrahim Effendi—a
very accomplished man, who introduced copper-plate
engraving into Constantinople in 1729. After describing his authorities, Major Rennell discusses the
itinerary measurements in different countries and the
various rates of travelling. He considers the caravan
journeys, the rate of camels, the days' journeys, as
given by the Arabian geographers Ibn Haukal and
Edrisi, and the mean rate of the marches of armies.
Then follow the descriptions of provinces, with critical
remarks on the geography and on the relative importance to be attached to the routes and accounts
of travellers.

This record of the construction of the comparative
geography of Western Asia was prepared for the use
of those who should in the future elaborate more improved systems of geography for the same tracts ; for
after so much labour and time had been employed in
collecting the materials, their use would have been
lost to future geographers had they only remained in
the mixed state in which they must necessarily exist
on a map. Hence the notes were kept together and

arranged, forming a complete record of the considerations and arguments for fixing each place on the maps and for the delineation of all the principal features.

It will be seen that the great work on the "Comparative Geography of Western Asia" was never quite completed. The "Geography of Herodotus," the illustrations of the history of the "Expedition of Cyrus," and the observations on the "Topography of the Plain of Troy" are portions of the work that was contemplated, each complete in itself. The treatise on the "Comparative Geography of Western Asia," with the accompanying maps, is the workshop, showing how the master worked with his geographical materials, and his method of building up the fair edifice which he left unfinished. It was a splendid conception, worthy of the great geographer, whose fame partly rests on the completed portions, especially on the "Geography of Herodotus." The latter work was not merely adapted for Rennell's own generation; it is of permanent use to geographical students. Sir Edward Bunbury, our greatest living authority on the subject, has recorded his opinion that Rennell's " Herodotus " is still of the greatest value.

CHAPTER VII.

In the end of the last century the attention of geographers was turned to the continent of Africa, as the region containing the largest extent of entirely unknown country and the most interesting problems to be solved. The absence of inland seas made its interior more inaccessible than that of the other continents, and the difficulties and expense involved in African travel made it almost impossible for private individuals, and seemed to render the formation of an association necessary to organise discovery.

Major Rennell's studies of Herodotus had already made him a very high authority on all matters relating to African geography. The whole of his second volume of the " Geography of Herodotus " was devoted to Africa, and the various questions are there treated with such clearness, perspicacity, and erudition that this second volume may be classed as the very best of Rennell's numerous works. He first investigates the question of the source of the Nile, of which Herodotus had very vague and contradictory notions: in one place supposing that the Nile rose to the westward of the greater Syrtis, and in another deriving it from the south. Ptolemy placed the sources far to the south of the equator, and he was followed by Edrisi.

Comparing the narrative of Bruce with that of Maillet, Major Rennell concluded that the Blue Nile and the Atbara were tributaries, and that the Bahr-el-abiad was the true Nile, with its source very far to the south-west of Abyssinia. He hesitated to place the source at so remote a distance as was adopted by Ptolemy or the Arabian geographers, but he thought that it was at least as far south as 6° N. His prophecy was that the true source of the Nile would be found when travellers made it their business to discover it, and not till then, because it probably lies far out of the touch of any caravan that visits the marts frequented by merchants who have intercourse with Europeans. He evidently looked forward to the time when the great Society which was the continuator and successor of the African Association, should send forth its Burtons and Spekes to solve the problem of ages and discover the sources of the Nile.

The sections on the delta of the Nile are most interesting, for here Major Rennell brings the great knowledge and experience he had acquired in surveying the Gangetic channels to bear on the questions connected with the changes in the Nile delta. He first discusses the Isthmus of Suez and the various statements of the ancients relating to the canal cut from the Pelusiac branch, through the bitter lakes, to the Gulf of Suez. He then enters upon the changes that have taken place in the delta since historic times, and on the causes leading to the choking-up of the Canopic and Pelusiac branches of the Nile. Rennell's general remarks on deltas and estuaries, on the positions of bars of rivers, and on the causes which lead to their formation, form an admirable exposition

of this part of the subject, and show the author's
mastery of a difficult but very important branch of
physical geography.

The oases of Egypt and Libya, which are called by
Abulfeda "the islands of the desert," were poetically
described in Thomson's "Seasons" as—

> "The tufted isles
> That verdant rise amid the Libyan wilds."

They had a special interest for Rennell, because they
were the stepping-stones of the caravan route from
Egypt to Fezzan, which he often had occasion to study
M. Poncet in 1698, and Browne in Rennell's own time,
described the oasis of Siwa, which contains the ruins
of the temple of Jupiter Ammon; and Browne observed
for latitude and longitude at Siout. It was by the
narratives of travellers who had visited these oases,
compared with Edrisi and Abulfeda, who give a
complete chain of distances from Cairo to Fezzan,
that Rennell established the position of Mouzourk,
which D'Anville had placed eighty miles too far north.
From the oases between Egypt and Fezzan, Rennell
conducts his readers to the gulfs between Carthage
and Cyrene, known to the ancients as the Greater
and Lesser Syrtis, and Lake Tritonis, including the
country of the lotus-eaters. Syrtis was the terror of
ancient mariners, because the coast was fully exposed
to the north and east winds, and there was great
difficulty and danger in working off a shallow lee
shore. Rennell was the first to suggest that the Lake
Tritonis of Herodotus was identical with the Lesser
Syrtis of later writers: a view in which he is supported by Rawlinson. He wrote a very detailed and

valuable dissertation on these Mediterranean gulfs
and on the tribes of Cyrenaica, who inhabited their
shores, including a discussion on the lotus of this
coast, described by Pliny, and, in Rennell's time, with
more scientific accuracy by Desfontaines, Dr. Shaw,
and Mungo Park.

But the most important sections in Rennell's
" African Geography of Herodotus " are those which
are devoted to the alleged circumnavigation of Africa by
Phœnician mariners under orders from Pharaoh Necho,
and to the Periplus of Hanno. In considering the story
of the circumnavigation, as told by Herodotus, Major
Rennell collected data to determine the rate of sailing
in ancient times, and their comparison gave an average
of only thirty-seven miles a day, always anchoring at
night. On an emergency, Nearchus was sometimes
under sail in very clear moonlight. He also considered
the effects of monsoons, trades, and currents, and the
time taken for sowing and reaping harvests, his
conclusion being that there was nothing to have
prevented the Phœnician sailors from performing
the circumnavigation of Africa in two years and a
half. The arguments on both sides are exhausted, as
Sir Edward Bunbury observed; and the absence of all
details respecting the alleged voyage precludes the
possibility of adding to them by further investigation.
Opinions will always continue to be divided: Gosselin,
Dr. Vincent, Mr. Cooley, and, of course, Sir George C.
Lewis, were incredulous. Rennell saw no reason for
disbelief, and his view was supported by Larcher,
Heeren, Rawlinson, and Grote. He constructed a
map to explain the circumnavigation of Africa,
showing the winds and currents.

The voyage of Hanno is also the subject of one of

Rennell's sections, and he gave a very elaborate and
valuable examination to the questions involved. He
also constructed maps to show the probable course of
Hanno's voyage, with currents, and Western Africa
according to Ptolemy. The "Periplus of Hanno" had
been translated by Falconer in 1797. Rennell was
the first to identify the "Southern Horn," or furthest
point of Hanno, with Sherborough Sound, just beyond
Sierra Leone. His argument is conclusive, and was
adopted by later editors. Sir Edward Bunbury says:—
" The merit of having first established the true view
of the question undoubtedly rests with the great
English hydrographer." *

These studies of African geography, extending
over many years, made Rennell a high authority
and a most invaluable coadjutor when the African
Association commenced its operations. There was a
tribe of Cyrenaica, called Nasamones, who are mentioned by Herodotus as having been great explorers.
We are told that a party of Nasamones made an expedition into the interior of Africa, with a view to
extending their discoveries beyond all former adven-
turers. They crossed the desert and were taken
prisoners by diminutive black men, who took them to
a city washed by a river abounding in crocodiles,
which may have been the Niger. Major Rennell
hails the Nasamones as the African Association of
their day.

For the African Association had come into existence while Rennell was still at work on his "Herodotus."
Its members, in announcing their plan, said that
" much of Asia, a still larger proportion of America

* Ancient Geography, I., 322.

and almost the whole of Africa, were unvisited and
unknown. The map of the interior of Africa is still
but a wide extended blank, on which the geographer,
on the authority of Edrisi and Leo Africanus, has
traced, with a hesitating hand, a few names of unexplored rivers and uncertain nations. The course of
the Niger, the places of its rise and termination, and
even its existence as a separate stream, are still
*undetermined. Nor has our knowledge of the Senegal
and Gambia rivers improved within the last half-
century, the falls of Felu on the first of these rivers,
and those of Baraconda on the last, being still the
limits of discovery. It is certain that while we continue ignorant of so large a portion of the globe, that
ignorance must be considered as a degree of reproach
upon the present age. Sensible of this stigma, and
desirous of rescuing the age from a charge of ignorance,
a few individuals, strongly impressed with a conviction
of the practicability and utility of thus enlarging the
fund of human knowledge, have formed the plan of
an Association for promoting the discovery of the
interior parts of Africa."

Accordingly, there was a meeting at the St. Alban's
Tavern on the 9th of June, 1788, when Sir Joseph
Banks was elected Secretary of the Association. The
Earl of Galloway, Lord Rawdon, Sir Adam Fergusson,
General Conway, and Messrs. Fordyce, Pulteney,
Beaufoy, and Stuart were present; the Bishop of
Llandaff being a member, but absent. It was agreed
that the subscription should be five guineas, and that
there should be a Committee to select persons to be
sent on expeditions of discovery, to form rules, and
superintend the expenditure. The Committee consisted of Sir Joseph Banks, Lord Rawdon, the Bishop

of Llandaff, Mr. Beaufoy, and Mr. Stuart. Major
Rennell was an honorary member of the Association.
The Bishop of Llandaff was the Honourable Shute
Barrington, afterwards Bishop of Durham for thirty-
five years. He had been a Fellow of Merton, and,
like his brother, Daines Barrington, was a keen
geographer. Henry Beaufoy, who was at first one
of the most active members of the Association, was
the son of a Quaker wine merchant in London. He
was in Parliament for many years, and for a short
time Secretary to the Board of Control. It was
Beaufoy who drew up the original plan of the
African Association and the Association's first report.
He died in 1795.* The Association soon numbered
137 members, including Earl Spencer, the Earl of
Wycombe, the Hon. J. Grenville, Gibbon the historian, the Speaker Addington, and Wm. Marsden.
Sir Joseph Banks held the office of secretary until
1797, when he was succeeded by Bryan Edwards, the
historian of the West Indies; and in 1801 Sir William
Young became secretary.

Eager to commence work with as little delay as
possible, the Committee at once selected Mr. Ledyard
and Mr. Lucas to penetrate into the interior of the
unknown continent. Ledyard had made a voyage
round the world with Captain Cook, as corporal of
marines. He received great kindness from Sir Joseph
Banks, who encouraged his enthusiasm for discovery.
Ledyard had conceived the idea of traversing the whole
extent of Arctic America—then unknown—from the

---

* Not to be confused with his namesake, Mark Beaufoy, the
magnetic observer, who was Colonel of the Tower Hamlets Militia.
Mark Beaufoy ascended Mont Blanc in 1787, six days later than
Saussure. Born in 1764, he survived till 1827.

Pacific to the Atlantic, and in order to reach the
threshold of his enterprise he resolved to make his
way by land to Kamschatka and cross Behring's Strait.
With only ten guineas in his purse, he crossed from
Dover to Ostend. Thence he found his way to Stockholm, and tried to cross the Gulf of Bothnia on the ice.
This being impracticable, he returned to Stockholm,
and set out on foot to the north, passing round the head
of the gulf, and reaching St. Petersburg, where he arrived without shoes or stockings. He drew a bill for
twenty pounds on Sir Joseph Banks, and was allowed
to join a detachment about to proceed with stores to
Yakutsk; and he found his way thence to Okzakoff.
Here he was suddenly seized, owing to some groundless
suspicion, and conveyed on a sledge, in the depth of
winter, to the Polish frontier, where he was turned
adrift. Worn with hardships, exhausted by disease,
nearly naked, he walked into the town of Königsberg.
Here he again ventured to draw for five pounds on the
President of the Royal Society. By this means he
reached London, and Sir Joseph Banks immediately
proposed to him to take service under the African
Association.

Sir Joseph sent Ledyard to Mr. Beaufoy, who was
" struck with the manliness of his person, the breadth
of his chest, the openness of his countenance, and the
inquietude of his eye." Spreading a map of Africa
before him, Mr. Beaufoy drew a line from Cairo to
Senaar, and thence westward to the supposed position
of the Niger, telling him that was his route. Ledyard
replied that he should think himself singularly fortunate to be entrusted with the adventure, and on being
asked when he would be ready to start, his answer
was : " To-morrow morning."

I

Mr. Lucas, the other explorer, had been sent to
Cadiz when a boy for education as a merchant. On
his voyage home his ship was captured by a Sallee
rover, and he was sent as a slave to Morocco. He
was redeemed after three years, and received the post
of Vice-Consul of Morocco from the Governor of
Gibraltar. After sixteen years, Mr. Lucas returned to
England, and accepted service under the African
Association.

Ledyard was instructed to traverse Africa from
east to west, in the latitude of the Niger; while Lucas
was to traverse the desert from Tripoli to Fezzan,
returning by the Gambia or the coast of Guinea.
These were very ambitious projects indeed; but the
continent was still a complete blank, and the difficulties were not fully understood. Ledyard left
England in June, 1788, arriving at Cairo in August,
where a fever put an end to the career of this
remarkable man. He was on the point of starting
with a caravan to Senaar. Lucas arrived at Tripoli
in October, 1788, and set out on his journey to
Mourzouk, the capital of Fezzan. He did not succeed in penetrating into the interior beyond this
point, returning to Tripoli in April, 1789. He,
however, collected a great deal of valuable information in Fezzan respecting the countries of the south.

In 1790 Major Rennell constructed a map of the
northern half of Africa for the African Association,
designed to illustrate all existing knowledge, and it
was accompanied by a very able and lucid memoir
on the materials for constructing such a map. He
begins by remarking that nothing can evince the low
state of African geography more than the fact of
M. D'Anville having had recourse to the works of

Ptolemy and Edrisi to construct the interior of
his map, published in 1749. Rennell was supplied
with numerous routes and itineraries, and he found
by long experience that one mile in eight must be
deducted to reduce the distances given to horizontal
measurement. But he was furnished with very few
fixed positions. In his map of 1790, Major Rennell
took the general outline, and the courses of the Nile,
Gambia, and Senegal, from D'Anville. By means of
caravan itineraries, the positions of Mourzouk, Agadez,
and Cashna were provisionally fixed, Cashna being
considered as the central kingdom of Northern Africa,
of which but few particulars were then known. The
Niger was believed to flow south of Timbuktu, and
was reported to be lost in the sands to the eastward.
Rennell found more difficulty in fixing the position of
Gonjah, which is the Conché of D'Anville and the
Gonge of Delisle ; the length of the journey of ninety-
seven days from Cashna rendering it very uncertain
where it should be placed. Timbuktu was placed on
the authority of Mr. Matra, British Consul at Morocco,
and of the reports of natives. The point of next importance was Bornou, for which there was a native
itinerary from Mourzouk. Rennell's map of Africa
was corrected in 1798, and again in 1802, when the
work of Houghton, Mungo Park, and other travellers
employed by the Association, were incorporated. It
thus formed a record of the discoveries, and of the
increase of geographical knowledge, made through the
exertions of the Association.

A detailed but exaggerated account of the inland
kingdom of Houssa had been received by the Association from an Arab, named Shabeni, and they were
anxious to ascertain the truth of his stories and to

I 2

discover the true course of the Niger. The Committee
therefore eagerly accepted the proffered services of
Major Houghton, formerly a captain in the 69th, and
who had since acted as fort-major at Gori. His plan
was to penetrate to the Niger by way of the Gambia.
His instructions were to visit Timbuktu and Houssa,
and to return by way of the desert, if possible.
Arriving at the entrance of the Gambia in November,
1790, he was kindly received at Medina by the King
of Wulli; but here a great misfortune befell him. A
fire destroyed not only his own house, with most of
his goods, but also the greater part of the town, while
his faithless interpreter decamped with his animals.
Nevertheless, he started, in May, 1791, in company
with a slave merchant; two asses carrying the wreck
of his property. Passing the former limit of European
discovery, he crossed the river Falemi and entered
the kingdom of Bondou, where he was badly received.
He, however, found a more hospitable welcome from
the neighbouring King of Bambouk, where a merchant
offered to conduct him on horseback to Timbuktu,
and to attend him on his return to the Gambia.
Major Houghton's last dispatches were dated July,
1791, and they enabled Major Rennell to prepare a
most interesting paper on the further elucidation of
African geography. A pencil note, dated September,
was received by Dr. Laidley, a resident at Pisania, on
the Gambia, in which Houghton reported that he had
been robbed of all his goods. He never returned;
and there was too much reason to fear that he was
stripped and left to die in the desert. The Association succeeded in obtaining a small provision for
Major Houghton's widow from the Government.

The next explorer who took service under the

African Association was a native of Hildesheim, in Germany, named Frederich Hornemann, who had studied at Gottingen, under Dr. Blumenbach and Professor Heeren, and had acquired a knowledge of Arabic. He was a young man, full of enthusiasm, with a strong constitution. Arriving at Cairo, he assumed the Eastern dress, and set out on his journey from Cairo to Fezzan in 1799. He arrived at Mourzouk, and his journals furnished ample materials for improving the maps of that region. They included an interesting account of his visit to the ruins of the Temple of Jupiter Ammon. Major Rennell prepared maps to illustrate Hornemann's journal, and wrote valuable geographical elucidations of his travels. Both Mr. Browne and Mr. Hornemann visited the oasis of Siwah and the ruins of the famous temple, and the latter made careful measurements. Rennell discusses fully the new information collected by Hornemann respecting the kingdom of Fezzan, as well as the improvements in the general geography of North Africa since the construction of his map, in 1790.

Mr. Browne's work was superior to that of the German traveller, because he fixed his positions by astronomical observations, while Hornemann's linguistic attainments enabled him to collect more information from the natives. Both possessed very great merits as travellers, in Major Rennell's judgment, and threw much light on the regions they described. Browne's materials consisted of 16° of latitude from Cairo to the capital of Darfur, corrected by observations for latitude and longitude ; while the information he obtained from natives reached to about the parallel of 8° N., and included reports touching the sources of the White Nile. Mr. Hornemann acquired the erroneous belief that the Joliba, or Niger, passed by the south of Darfur into the White Nile. Rennell thought it probable that the Niger terminated, by evaporation, in the country of Wangara, to the westward of Bornou. Mr. Hornemann collected a good deal of information respecting the people and countries south and west of Fezzan, including Houssa and Timbuktu. Hornemann set out from Mourzouk with the object of reaching Bornou. Years passed away, many inquiries were instituted, and rumours of Yusuf—the name under which Hornemann travelled—were occasionally received; but no authentic intelligence was ever obtained, and his fate is unknown.

Meanwhile, the Association was occupied in seeking for a successor to Major Houghton, who would undertake a similar journey from the Gambia to the Niger. The Committee was most fortunate in its selection of Mungo Park. Born in 1771, in a cottage in the glen of Yarrow, where his father was a small farmer with a large family, the future explorer was sent to Selkirk Grammar School. He was next apprenticed to a surgeon at Selkirk, and went thence to Edinburgh University. In 1792 he went out to India as a surgeon in the East India Company's sea service. His first voyage was to Sumatra, and on his return he offered his services to the African Association; and they were promptly accepted. Sailing from England in May, 1795, Mungo Park, who had just reached the age of twenty-four, proceeded up the River Gambia. He was very hospitably received by Dr. Laidley at the factory of Pisania, where he passed several months. His instructions were to reach the River Niger, and to ascertain its source, its course, and, if possible, its termination; and to visit Timbuktu and the towns of the Houssa country. After acquiring some knowledge of the Mandingo language, Mungo Park set out from Pisania in December, 1795. Passing through Medina, where he was civilly received by the King of Wulli, he left the Gambia, and took a course towards the Senegal River, over the Bondou country, inhabited by the Moorish race of Fulahs. Crossing the Senegal, he adopted a circuitous route by way of Ludamar, an Arab district, to Bambarra, on the Niger.

It was in this part of the country that Park fell in with the lotus—a plant to which Major Rennell had devoted much attention and research in the part of his work on Herodotus relating to the Lotophagi, on the shores of the Gulf of Syrtis. In riding along, Mungo Park and his followers came upon two negroes who had come to gather what they called *tomberongs*. These proved to be small farinaceous berries, of a yellow colour and delicious taste: the fruit of *Rhamnus lotus* of Linnæus. Mungo Park gives a very interesting account of this famous plant: "The berries are much esteemed by the natives, who convert them into a sort of bread, by exposing them for some days to the sun, and afterwards pounding them gently in a wooden mortar until the farinaceous part of the berry is separated from the stone. This meal is then mixed with a little water and formed into cakes, which, when dried in the sun, resemble in colour and flavour the sweetest gingerbread. The stones are afterwards put into a vessel of water and shaken about, so as to separate the meal which may still adhere to them. This communicates a sweet and agreeable taste to the water, and, with the addition of a little powdered millet, forms a pleasant gruel, which is the common breakfast in many parts of Ludamar during February and March. The fruit is collected by spreading a cloth upon the ground and beating the branches with a stick." The lotus is very common in all the kingdoms visited by Mungo Park, especially in Ludamar and the northern parts of Bambarra; and Major Rennell looked upon its identification as one of the most interesting collateral results of his young friend's journey. The plant had previously been figured by Desfontaines in the "Memoires de l'Academie Royale des Sciences," 1788. Park observes that there can be little doubt of its being the lotus mentioned by Pliny as the food of Libyan Lotophagi.

Park was detained and imprisoned by Ali, the Chief of Ludamar, on the borders of the Great Desert, and escaped with difficulty, after a long detention. At length the gallant explorer, in July, 1796, reached Sego, the capital of Bambarra, where he first saw "the great object of his mission—the long-sought-for majestic Niger—glittering to the morning sun, as broad as the Thames at Westminster, and flowing slowly to the eastward." It was when refused permission to cross the ferry to Sego, denied hospitality by all the inhabitants of a neighbouring village, and about to pass the night under a tree, hungry and tired, that the woman took pity on him, whose people sang the extempore song we all know so well:—

"The winds roared and the rains fell,
   The poor white man, faint and weary, came and sat under
     our tree.
   He has no mother to bring him milk,
   No wife to grind his corn ;
   Let us pity the white man: no mother has he."

The King of Bambarra dismissed Park with a present
of a bag of cowries, and he pursued his journey to the
eastward. He reached the town of Silla, where he
determined to retrace his steps, for the tropical rains
were setting in, and he seemed to be getting into the
country of merciless and fanatical Moors. He was
informed that it was fourteen days' journey from Silla
to Kabra on the Niger, which is but one day from
Timbuktu. Eleven days' journey down the stream
the river passes Houssa, but all the natives seemed to
be ignorant of the further course of the Niger and of its
mouth. Houssa was reported to be a great mart for
Moorish commerce. After enduring many insults and
terrible hardships, Park at length succeeded in joining
a caravan and in reaching Pisania, where he was
warmly welcomed by Dr. Laidley, who had given
him up for lost. He reached England in December,
1797.

The excitement in London was so great, and the
eagerness to know the details of Park's journey so
importunate, that the African Association issued a
preliminary report, written by the secretary, Mr. Bryan
Edwards. Major Rennell worked out the routes of
Mungo Park with great care, and his geographical
illustrations were published, with a map of Park's
route, which was afterwards used to illustrate his
book. Rennell considered that the journey of Mungo
Park had brought to the knowledge of his generation
more important facts respecting the geography of
Western Africa than had been collected by any
former traveller. By pointing out the positions of
the sources of the Senegal, Gambia, and Niger, he
showed where to look for the elevated parts of the
country and for the water-partings between the

Gambia and Niger, as well as the boundary between
Moors and Negroes, and between the fertile country
and the desert.

Rennell traced the history of opinion respecting
the course of the Niger from the earliest times.
Herodotus described the river as flowing from west to
east, dividing Africa as the Danube divides Europe.
He held the Niger to be a remote branch of the Nile.
Ptolemy describes the Niger as a separate stream
from the Gambia and Senegal, and as extending from
west to east over half the breadth of Africa. But
Edrisi took the Niger westward into the Atlantic,
and he was followed by Abulfeda. The early Portuguese held the same opinion, thinking that the Niger
was merely the upper course of the Senegal. De
Lisle's map of Africa (1707) gives the Niger a direct
course through Africa, rising in Bornou, and ter-
minating in the Senegal. But in his subsequent
editions of 1722 and 1727 this blunder is corrected,
and the course of the Niger is turned east towards
Bornou, where it terminates in a lake. The cause of
this correction may be traced to information from
Frenchmen settled on the Senegal. D'Anville followed
the later editions of De Lisle.

Major Rennell then proceeds to discuss the oro-
graphy of the region explored by Mungo Park.
According to Leo Africanus, the country of Melli is
bordered on the south by mountains, and these
must be nearly in the same parallel as the mountains of Kong, seen by Park. Mr. Beaufoy was
also informed that the countries south of the
Niger were mountainous and woody. The evidence
respecting the Kong mountains seemed to preclude
the idea that the Niger turned to the south and

emptied itself into the Gulf of Guinea. The other alternative was that it continued to flow to the east, finally losing itself in a lake in the far interior, called Wangora by Edrisi. But this was merely a provisional theory, which best fitted the information actually obtained. Major Rennell was not in the least wedded to it. He always said that the place and mode of the termination of the Niger were not exactly known; and long afterwards, in a letter to Sir Francis Beaufort, he wrote :—" If Denham is successful, we may know what becomes of the Niger." The elucidation of Mungo Park's routes, and the construction of the map of his travels, was a most important service, for which geographers are indebted to the learning, skill, and industry of Major Rennell.

The remarkable success of his first journey induced the Government to employ Park to complete the discovery of the course of the Niger. He received a captain's commission, with his brother-in-law, Anderson, as his lieutenant, and he was to be accompanied by forty-five European soldiers, besides natives. Sailing from England in January, 1805, all was ready for the march by April ; but it was a fatal mistake to employ European soldiers, and disease broke out within the first week of leaving Pisania. Many died ; and it was not until August that the Niger was reached. The party was embarked in canoes, but it had dwindled down to five men, besides Park and Anderson ; and in October Anderson died. Park was resolved to discover the mouth of the Niger, or die in the attempt. He sailed over a thousand miles down the river, but below Yuri he encountered rapids, while both shores were lined with hostile natives. The little vessel of the explorers was hopelessly

jammed in a cleft of the rocks; all efforts to free
her proved unavailing, and finally Park and his three
surviving companions jumped overboard, and were
drowned. Thus did this great traveller meet an
heroic death in the midst of his discoveries. He set
a glorious example of devotion to duty, of daunt-
less courage, and unswerving resolution; and his
name will ever be held in honour by succeeding
generations.

In 1820 the English Government resolved to equip
an expedition with the object of penetrating southwards from Fezzan in the track of poor Hornemann.
It consisted of Lieutenant Clapperton, R.N., Major
Denham, and Dr. Oudney, with servants and an Arab
escort. They left Mourzouk, the capital of Fezzan, in
November, 1822, and commenced the journey across
the sandy desert. By February, 1823, they had
reached the shores of Lake Chad, and on the 17th
they arrived at Kuka, the capital of the Bornu State.
Clapperton and Oudney, after a delay of many
months, set out for Kano and the Houssa State.
Dr. Oudney died on the road in January, 1824, but
Clapperton pressed onwards, passed through Kano,
and arrived at Sokoto, the capital of a new State
recently founded by a Muslim people called Fulahs.
He hoped to continue his journey to Yuri, and
complete the work left unfinished by Park. But the
Sultan of Sokoto raised obstacles, and Clapperton was
obliged to abandon his plan, and return to Bornu and
Fezzan, reaching Tripoli early in 1825. It was then
resolved to send Clapperton to Sokoto, in order to
establish direct intercourse with the Sultan, and it was
decided that he should proceed to his destination by
landing in the Gulf of Benin. Eventually, Clapperton

landed at Badagri, near Lagos, in December, 1825, but all the members of his party were obliged to remain behind, owing to attacks of fever, and he proceeded with only one companion, his servant, Richard Lander. They reached the Niger at the point where Park had lost his life, and proceeded thence to Sokoto. Here Captain Clapperton succumbed to fever, dying in April, 1827. Lander resolved to carry out his late master's intention of tracing the Niger to its mouth; but first he went down to the coast at Badagri and brought home his master's journal. The Government accepted his proposal to complete the work, and he landed at Badagri again, with his brother John as a companion, in March, 1830. Having reached Yuri, on the Niger, the brothers obtained two canoes, and commenced the descent of the river. They passed a range of mountains, which has since been named after Major Rennell, and entered the wide reach at the confluence of the Benue. Continuing the descent, they came to the sea on the 24th of November, 1830.

This discovery immediately led to the inauguration of enterprises with the object of opening the trade of the Niger region. Macgregor Laird fitted out two steamers at Liverpool, and went out, accompanied by Mr. Lander and Lieutenant Bird Allen, an accomplished naval surveyor. They were engaged in exploring the Niger delta for several weeks, and in the following season Lander ascended the Benue for a considerable distance. But the great majority of the men died of fever, and Richard Lander, having been shot, was conveyed to Fernando Po, where he expired in 1832.

All this work was proceeding during the last ten

years of Major Rennell's life. There can be no doubt that the Government expeditions were due to the initiative of the African Association, and that the second enterprise of Mungo Park, as well as the subsequent expeditions of Denham, Clapperton, and Lander, all resulted from the first journey of Park, which was conceived, organized, and despatched by the Association. Rennell was always deeply interested in the question of the outlet of the Niger. He had carefully studied every authority on the subject, from Herodotus to Mungo Park. The latter believed that the Niger flowed into the Congo; but the weight of evidence was that mountains intervened between the Niger and the Gulf of Guinea, and that, consequently, the great river flowed eastward until it was lost in some central lake. Major Rennell adopted this view provisionally, but without any strong bias, as is clearly proved by his private correspondence on the subject.

It is true that as early as 1816 Mr. James M'Queen had started a theory—which eventually proved to be correct—that the Niger entered the ocean in the Bight of Benin, and in 1821 he brought out a fuller treatise on the subject. But his evidence consisted of stories told him by negroes on an estate in the West Indies, and other equally reliable informants. It was little more than a lucky guess, and not sufficient to alter the opinions of serious and thoughtful students. M'Queen misrepresented the words of Rennell, making him say that the Niger disappeared in wastes of sand, or became evaporated in swamps under the heat of a tropical sun. Major Rennell said nothing of the kind. He always used the word lake, referring to the large central lake of Edrisi, and others, as the possible receiver of the Niger, if the Kong Mountains separated it from the sea.*

In his old age Major Rennell watched for the solution of the Niger question with ever-increasing interest. He expected much from Major Denham. When the news of Clapperton's journey from Lake Chad to Sokoto reached England, the great geographer must have seen that the Niger did not flow eastward, as he had supposed; but he did not live long enough to receive the full solution of the problem. Major Rennell died in the very month that Richard Lander reached the mouth of the Niger.

African geography owes much to the elucidations of Rennell and to his indefatigable research. His admirable disquisition on the delta of the Nile is a masterly contribution to physical science. His identification of Lake Tritonis of Herodotus with the Lesser Syrtis of later writers has been generally accepted. His disquisition on the alleged circumnavigation of Africa by the Phœnicians has exhausted that question, and he arranged all the facts—historical and scientific —with such lucidity, that any one may form a judgment on them with the assurance that all the premises are before him. His examination of the Periplus of Hanno is a masterpiece of critical reasoning, supported by every consideration that bears on the subject.

---

* Mr. Joseph Thomson, in his " Life of Mungo Park," speaks with exaggerated admiration of M'Queen's imaginary geography, and sneers at Major Rennell as "an arm-chair geographer," and as "the man with one idea." It is a strange misconception to speak of the surveyor of Bengal, who worked in the field for fifteen years, and was broken down by wounds and fevers, as an "arm-chair geographer;" and a still stranger misconception to refer to the savant who, of all others, was most open to conviction, and the most tolerant of adverse opinions, as "the man with one idea."

Rennell's hydrographic knowledge here throws light
on points which would otherwise be obscure, and his
furthest point reached by Hanno is now generally
accepted. The map of Northern Africa, prepared for
the use of the Association, marks an important era in
the progress of discovery. It is the result of immense
research, combined with sagacious and thoughtful
reasoning, and is decidedly in advance of D'Anville.

Rennell's latest African work embraced his elucidations of the reports of explorers employed by the
Association, and his maps to illustrate their travels.
With regard to the journals of Hornemann, the great
geographer compared his statements and measurements with the work of Browne and with the accounts
of the Arabian geographers. He was thus enabled to
improve the delineation of the region between Egypt
and Fezzan, with its interesting oases, and also to
correct the positions of Mourzouk and other places
in Fezzan. Rennell often had to work with very
inadequate materials, but he always had a generous
word for the shortcomings of explorers. Speaking of
Hornemann, he wrote :—" Very great allowances must
be made for the situation in which he was placed, and
the difficulty he had in supporting the character he
had assumed."

Perhaps Rennell's greatest service to African geography was the way in which he worked up the rough
notes of Mungo Park, examining his daily routes with
the greatest care, comparing them critically with the
work of all previous travellers and cartographers, and
bringing all the materials into harmony so far as
was possible. These labours enabled him to construct a map of the discoveries of Mungo Park, which
served to illustrate his travels. Rennell's map and

geographical memoir were valuable additions to Park's volume.

The African researches and labours of the great geographer extended over many years, and the deep and intelligent interest he took in the progress of discovery only ended with his life.

CHAPTER VIII.

Rennell's Current.

MAJOR RENNELL was, before all things, a sailor. He never forgot that he had been a midshipman. He showed this in the deep interest he always took in naval affairs, in the friendships he retained and in the new friendships he made, in his correspondence, and, above all, in the enormous amount of labour and trouble he devoted to the study of winds and currents: to the branch of his science which is now called oceanography. His numerous naval friends furnished him with a great mass of materials from their logs and note-books, and he prosecuted his inquiries with untiring zeal during a long course of years. It was in about 1810 that he began to reduce his collections to one general system, receiving much assistance from Mr. John Purdy, the eminent hydrographer.

The first chart published by Major Rennell was of the banks and currents at the Lagullas, in South Africa, dated November, 1778, and inscribed to Sir George Wombwell, the Chairman of Directors of the East India Company. His accompanying memoir was printed, forty years afterwards, in Purdy's "Oriental Navigation." In 1797 the Court of Directors resolved to have surveys executed of the Isle of Wight, the Isle of Thanet, and other parts of the coast of England, "for the purpose of ascertaining the most eligible situation for a depôt for receiving, training, and disciplining recruits for India." Major Rennell was requested to give his assistance to the Committee of Shipping, with a view to the execution of these surveys.   He had previously, in 1793, presented the Court with a chart of Mount's Bay and the adjacent coasts.   At this time, also, he was in correspondence on naval and other subjects with his constant friend Lord Spencer, who was First Lord of the Admiralty from 1794 to 1800.   Among Major Rennell's letters preserved at Spencer House, there is one, dated 1795, in which he discusses the construction of floating batteries in much detail.   In 1799 the question of fire occupied Lord Spencer's attention, in consequence of an accident on board some man-of-war, and we find his lordship discussing details of fire quarters with Major Rennell down to the special beat of the drum.   Other letters are on subjects not relating to naval matters.   One discusses the courses and positions of waters in the Hudson's Bay Company's territory and the routes across the North American Continent.   Another gives a graphic account of a case of suttee witnessed by Major and Mrs. Rennell.

In 1792 we find Major Rennell taking a deep interest in the dawn of the science of geology. Dr. John Hunter, the eminent physician, had written some observations and reflections on geology, intended to serve as an introduction to the catalogue of his collection of fossils.*   Dr. Hunter had sent his

* Printed in 1859 (p. 58) with Major Rennell's letter. I am indebted to Sir William Flower for calling my attention to Dr. Hunter's interesting pamphlet.

J 2

manuscript to Major Rennell for his remarks.   In
returning it, Major Rennell wrote:—" I have kept
your manuscript longer than I ought to have done,
but it was that I might read it over more than once,
and have such intervals between the readings that the
arrangement might in some measure appear new to
me: a practice I follow with regard to my own
compositions, for reasons that you, no doubt, have
long ago thought of.   I have been very much instructed and interested throughout, and shall be glad
to see the remainder of it when completed.   I have
indeed read it three times, by way of first reading,
commitment, and report."   He then arranged a
meeting with Dr. Hunter to discuss various points ;
but meanwhile, he offered one observation on the
desirability of respecting prejudices, which is interesting.   He says: " At page 3 you have used the term
"many thousands of centuries,' which brings us almost
to the *yogues* of the Hindus.   Now, although I have
no quarrel with any opinions relating to the antiquity
of the globe, yet there exists a description of persons
very numerous and very respectable in every point,
except their pardonable superstitions, who will dislike
any mention of a specific period that ascends beyond
6,000 years.   I would therefore, with submission,
qualify the expression by saying 'many thousand
years,' instead of ' centuries.'"   Rennell lived to see
the science of geology, which had its earliest dawn in
this country in the pamphlet of John Hunter, become
a recognised branch of research, with a Society of its
own, and a series of rapidly succeeding discoveries
developed by numerous ardent and enthusiastic
workers.

Rennell was himself the founder of another branch

of the science of geography, which has since been called oceanography. His current charts and memoirs were completed, although they were not published in his life-time. He himself pointed out that the work could not have been usefully undertaken at any earlier period; because until the method of ascertaining longitude by chronometer had been invented and put in practice, the necessary positions could not have been fixed, and current charts could not have been drawn with any approach to accuracy.

The discovery of a method of finding the longitude at sea had long been a desideratum. In 1714 an Act of Parliament was passed, offering rewards from ten thousand pounds to twenty thousand pounds for the discovery, and creating a Board of Longitude, which included the Astronomer-Royal, the President of the Royal Society, and the Master of the Trinity House. At length, in 1726, a time-keeper, constructed by Mr. Harrison, was the means of correcting an error of a degree and a half between London and Lisbon. Harrison made still more accurate watches between 1739 and 1760. In the year 1761, his son, William Harrison, embarked for Jamaica with his father's most improved watch, and on his arrival it was only one and a quarter minute from the true longitude. Eventually, but after long and unjust delay, the Harrisons received the reward of twenty thousand pounds, and from that time the system of finding the longitude at sea by chronometer was established. Neville Maskelyne, the Astronomer-Royal, had commenced the issue of a "Nautical Almanack" in 1767, containing tables of declination and distances of the moon from the sun and fixed stars, computed for the meridian of Greenwich, expressly designed for finding the longitude at sea. The "Nautical Almanack" has been published annually ever since, and has been much enlarged.

These improvements in the science of navigation enabled an idea of the direction and force of ocean currents to be formed in some detail. Progress in the knowledge of ocean currents has consequently been very rapid since the invention of chronometers. The use of such knowledge is sufficiently obvious; for in whatsoever direction a portion of the ocean may move, it must either favour or impede a ship's course. A knowledge of the currents will, therefore, enable the navigator so to shape his course as to avoid delay or danger. Major Rennell supplies examples:—" For instance, a just idea of the nature of the equatorial current would prevent a commander from crossing the equator too far to the westward in the southern passage, particularly at the season when the S.E. trade wind is very far southerly, and when also the current runs so strongly to the westward in the neighbourhood of Cape St. Roque as to hazard his being driven to leeward of it. Again, a winter passage round the Cape of Good Hope to the westward may depend on keeping in the stream of the Lagullas current. So that it was very truly said before chronometers came into use, by Sir Charles Blagden, in writing of the Gulf Stream in 1778, that hitherto the difficulty of ascertaining currents is well known to be one of the greatest defects in the present state of navigation."

Major Rennell's work was first to collect materials for illustrating and explaining the subject of the currents of the ocean generally, more particularly for those in the Atlantic and Indian Oceans, and then to form a system in conformity with the observations. His personal knowledge and his professional experience as a sailor furnished him with the means of appreciating the value of the materials at his disposal, as well as of adding, in some instances, the results of inquiries and observations made on the spot. " The formation of a great number of facts into a system," wrote Rennell, " may prove of use in impressing those facts on the mind more strongly than would be the case it they were left to operate independently of each other ; for a fact often makes less impression when standing naked and alone than when it makes part of a system, which operates like a band to keep the parts together · in their proper places, when they explain and illustrate each other."

His volume on the winds and currents is confined to the principal streams in the North and South Atlantic Ocean, and those which pass between the Indian and South Atlantic Ocean round the Cape. It also treats of the regions of the trade winds in the two Atlantics, showing the changes that take place in the different parallels and seasons. But although the scope of Major Rennell's volume was thus confined, he had acquired by diligent inquiry an extensive knowledge of the winds and currents in the Pacific Ocean and other parts of the world.

As the winds are to be regarded as the prime movers of the currents of the ocean, Major Rennell commences by explaining the directions of the trade winds and monsoons, and the changes to which they are subject. Although trade winds are denominated N.E. and S.E., yet they both vary at different seasons and in different parallels, for they have at all seasons a direct tendency to blow towards the place

of the sun, or less wide of it. Thus the N.E. trade
is more northerly when the sun is in the southern
signs, and the S.E. more southerly in the opposite
season. Both trades also, when free from monsoon
influence, blow more southerly and northerly in
regions adjoining to the old continents than towards
the middle of the ocean, so that as we recede from the
coast the wind gradually becomes more and more
easterly, and finally almost an easterly wind.

After discussing the trade winds and monsoons,
Major Rennell devotes a section to a general view of
the system of currents in the Atlantic. The first
in his system is the South Atlantic current. A
stream off the Cape of Good Hope which makes its
way round the Cape and the Bank of Lagullas is a
part of the Lagullas current. By the time that it
reaches the mouth of the Congo it has become a
powerful and extensive stream. It turns to the westward along the equator with the bend of the land,
while the Guinea current from the North Atlantic
passes within it to the Bight of Benin. The South
Atlantic current continues its course along both sides
of the equator, becoming an equatorial current, and
forming a wide and complete bar across the narrow part
of the Atlantic between Guinea and Brazil. Between
the two continents it sends off a large branch to
the N.W., while the main stream turns W.S.W
towards Cape St. Roque, where it subsides, part
going towards the West Indies, the rest along the
coast of Brazil and Patagonia. The N.W. equatorial
stream enters the Caribbean Sea through the passages
between the islands.

Thus the Gulf of Mexico is supplied with the great
head of water which makes it the reservoir of the

Gulf Stream. Rennell looked upon the Gulf Stream in the nature of an immense river descending from a higher level into a plain. Commencing at the head of the Florida Strait, it pursues its way, not far from the coast of the United States, to Cape Hatteras, where the coast turns more to the left, while the stream is gradually deviated more and more to the eastward, finally pointing E.N.E., until it touches the parallel of 44° 30' N. in about 43° W., or midway between New York and Cape Finisterre. This idea of a river in the ocean was thus due to the imaginative mind of Rennell, although it was adopted and amplified by Maury in his "Physical Geography of the Sea." Rennell held that the Gulf Stream terminated on the western side of the Azores, and that some of its waters thence passed southwards, carrying with it a great deposit of the sargasso, or gulf weed. Another portion is propelled by westerly winds towards the Bay of Biscay and the coast of Morocco. Rennell was furnished with no less than ten different traverses, or crossings, of the Gulf Stream, in which the temperatures were registered; and he was also in possession of a very considerable number of examples of the direction and rate of the stream.

Major Rennell next proceeds to investigate the Arctic current, which, coming from Greenland, strikes the Gulf Stream to the east of the Great Bank of Newfoundland, in about 44° N., and between the meridians of 44° and 47° W. But his observations respecting the Arctic current were by no means so full and complete as those relating to the Gulf Stream; and the former was first properly examined and explained by Mr. Redfield in 1838, and by the operations of the United States Coast Survey. It was found that the

Gulf Stream closes with the land at Cape Hatteras, and in its subsequent progress to the N.N.E. maintains a nearly perpendicular wall of warm water, in contact with the cold Arctic current flowing south. Mr. Findlay has since pointed out that the fact of the non-blending of the warm tropical with the cold Arctic waters, when in juxtaposition, gives great weight to a suggestion made by Mr. Redfield that the Arctic current passes under the Gulf Stream. The bifurcation of the Gulf Stream discovered by Professor Bache off Cape Hatteras was also unknown to Rennell.

Few circumstances of the kind have occasioned more surprise than the distance to which the warm water of the Gulf Stream is carried in its progress through the Atlantic Ocean, and the vast extent of the warm water. Rennell was enabled to prepare two Tables of the Temperatures through the whole extent of the Stream, from the Gulf of Mexico to the Azores. His calculation was that the stream, in running through about 3,060 miles, and altering its latitude eighteen degrees, diminishes its temperature about thirteen and a half degrees, and may be supposed to occupy from seventy-six to seventy-eight days on its passage.

In November, 1776, Dr. Franklin, on a voyage from Philadelphia to France, observed that he was never out of the warm water of the Gulf Stream, showing that it occasionally extends to the coast of Europe. The distance from the island of Corvo to the head of the Bay of Biscay is about 1,150 miles and it was found by three different observers that the velocity of the stream, in August and September of different years, was from thirty to thirty-three miles during the twenty-four hours, between the meridians of 41° and 45° W. At about 40° the stream begins to turn towards the south of east, and gradually round to S.E. to Corvo and Flores, while its velocity is much reduced. Rennell concluded that the Gulf Stream terminated at the Azores in ordinary years, and that, in the instance noted by Dr. Franklin, the body of the stream had a more than usual degree of velocity, making it continue on the course through the Atlantic instead of turning to the south. Nothing higher, in ocean temperatures, had been observed between the Azores and the shores of Europe since Dr. Franklin's voyage in 1776, until Captain Sabine found a similar result in 1821.

Subsequent research proved that the Gulf Stream does not stop at the Azores, though it ceases to be a weed-bearing current. A portion is diverted to the N.E., and the rest passes towards Madeira, and strikes on the shores of Morocco.

Major Rennell defines the North African and Guinea current as caused by an accumulation of water towards the eastern and south-eastern parts of the tract to the northward of the Gulf Stream, which runs off to the S.E. Originating between 43° and 53° N., the stream called the North African and Guinea current runs to the south-east and south as far as the coast of Guinea. It is augmented in its way by drift currents, so as to become a powerful stream. A ship sailing southward from England will generally find a south-easterly current between the Bay of Biscay and Madeira, which becomes east in the parallel of the Strait of Gibraltar. Its rate is from twelve to twenty miles a day. Rennell observes that it is this current which often occasions unwary navigators to

be thrown ashore, even in fair weather, on the coast
of the Sahara. He found that the study of the North
African stream was intricate, as it has various ramifications which take different directions, though all of
them, except those which point towards the entrance
of the Strait of Gibraltar to supply its current, point
in some degree southward. From about sixty miles
N.W. of Cape St. Vincent to Cape Cantin the currents
from every quarter point towards the Straits' mouth
as to the pipe of a funnel, of which the reservoir is
the semicircular space between Cape St. Vincent and
Cape Cantin. But it does not seem to have been
suspected, in Rennell's time, that this influence extends
very far into the Atlantic in different directions, and
especially during the summer, when the evaporation
of the Mediterranean, which these currents are
supposed to displace, must be greatly increased.

From the Straits' mouth to Cape Bojador, the
motion of the sea, for a distance of a hundred leagues
from shore, points obliquely towards the land, and
much the same state of things prevails as far as Cape
Blanco, at the rate of half-a-mile per hour. At times
this inset is much stronger than at others. M. D'Apres
reported that vessels had made the land of Africa,
when they expected to have made Teneriffe. From
whatsoever cause it may arise, the effect of this drift
towards the shore has caused numberless shipwrecks.
Rennell adds:—" Perhaps no other current in the ocean
has ever produced so much misery to sailors and
passengers. It is the operation of these currents that
has placed from time to time a number of shipwrecked
captives of all nations in the hands of the barbarous
tribes on the western edge of the great African
Desert, who sell the survivors to the scarcely less

barbarous people of Morocco. These accidents appear
to owe their immediate cause to the wish to save a
little distance by making a straighter course, and
keeping nearer to the shore. In reality, their progress
would on the whole be much greater if they sailed
more at large until they had passed Cape Blanco ; for
the winds are more steady outside, and the current is
directly in their favour, which is not always the case
nearer the land. The coast of the desert, being very
low, cannot be seen at a distance, so that it happens
that a ship which, by the gradual operation of the
current on her course, has been carried towards the
shore, arrives so near it by the evening that, if it had
been of ordinary height, it might have been seen and
avoided. Continuing her course during the night,
she runs on shore before those on board are aware of
the vicinity of the land or of being in soundings.
This seems to have been the case of Admiral Keppel
in the voyage to Gori, during the seven years' war.
The *Lichfield*, of fifty guns, was lost, and her crew
made slaves, while the rest of the squadron had a
narrow escape."

Rennell makes this current follow the whole West
Coast of Africa to the Bight of Biafra as the Guinea
current ; but before arriving at the Cape Verdes, it
would trend off to the S.S.W. and S.W., the direction
of the trade wind in this part. Multiplied subsequent
observations have shown that such is the fact. Mr.
Findlay, in 1853, pointed out an origin of the Guinea
current not hitherto suspected. The Guinea current,
he held, is a portion of the equatorial current itself,
here deflected, and coming from the westward between
5° and 10° N. It is, in fact, a warm current, while the
North African current is comparatively cold. The

main body of the North African current turns to the
S.S.W. and S.W., and then to the west, joining the
equatorial stream, and the circulation is thus around
the Sargasso Sea, which lies almost across the
Atlantic, in the parallel of 30° N. The circulation
of the surface waters of the Atlantic round this central
space is thus satisfactorily explained. Maury illus-
trates the matter in his graphic way:—"If a bit of
cork, or any floating surface, be put into a basin, and
a circular motion be given to the water, all the light
substances will be found crowding together near the
centre of the pool, where there is the least motion.
Just such a basin is the Atlantic Ocean to the Gulf
Stream, and the Sargasso Sea is the centre of the
whirl. Columbus first found this weedy sea in his
voyage of discovery, and there it has remained to this
day, moving up and down, and changing its position,
like the calms of Cancer, according to the seasons, the
storms, and the winds."

The increase of waters brought into the Atlantic
by the Arctic current is compensated for by the outlet
along the Norwegian coast.

It was on reaching the Cape Verde Islands with
his North African current that Rennell made his
mistake. Instead of assuming, as is the case, that it
turned west into the equatorial stream, he continued
its direction south and south-east, supposing that the
North African and Guinea currents were one stream.
On this question he had studied the journals of
Captain Cook, Admiral Bligh, Captains Krusenstern,
and Lisiansky. He considered that it had been
satisfactorily ascertained that this Guinea current
was four degrees in breadth, because Captain Cook's
two tracks of 1772 and 1776 through it were at that

distance from each other. It is now well established that the Guinea current is not a continuation of the North African current, but that it is the eastern portion of a counter equatorial current. Mr. Buchanan found the easterly current to be very strong, especially inshore, along the coast of Guinea; and north of the island of St. Thomè there are large areas of river water from the Niger on the surface, with the Guinea current water below. It was found that this Guinea current experienced marked variations with the seasons. But this counter equatorial current still requires investigation. Mr. Buchanan says that " it is particularly interesting, and its dynamics obscure. Its range is very superficial, and its physical conditions can be studied without the elaborate and costly equipment required for the research of oceanic depths."

Major Rennell did not succeed in collecting information respecting the northern flow of the waters of the Gulf Stream. Crossing the fortieth degree of latitude, the warm water spreads itself out over the cold waters around, and, in Maury's words, " covers the ocean with a mantle of warmth that serves so much to mitigate in Europe the rigours of winter. Moving more slowly, but dispensing its genial influences more freely, it finally meets the British Isles, and flows thence into the Polar Sea, between Spitzbergen and Novaya Zemlya." Nor was it possible for Rennell to describe the Arctic currents with any precision, because the necessary observations had not then been taken. He had the results of work done by Captain Beaufort off Newfoundland in 1808, those of Sir John Duckworth, and the observations of Captain Parry on the currents in Davis Strait and along the coast of Labrador.

The current now known as Rennell's Current was brought to the notice of the Royal Society in papers read on the 6th of June, 1793, and April 13th, 1815. The title of the first paper was " Observations on a Current that often prevails to the westward of Scilly, endangering the safety of ships that approach the English Channel, and now known by the name of Rennell's Current."

It had long been well known to seamen that ships in coming from the Atlantic and steering a course for the Bristol Channel in a parallel somewhat to the south of the Scillies often find themselves to the north of those islands. This extraordinary error has passed for the effect either of bad steerage, bad observations for latitude, or the indraught of the Bristol Channel. But none of these reasons account for it satisfactorily. Admitting that there is an indraught at times, it cannot be supposed to extend to Scilly, and the case has happened in weather most favourable for navigation and for taking observations. The consequences of this deviation from the intended track have very often been fatal. There was the loss of the *Nancy* packet, in Rennell's own time. But the most memorable instance was that of Sir Cloudesley Shovel, when he was returning from the Mediterranean with twenty sail of vessels. With his flag on board the *Association*, Admiral Sir Cloudesley Shovel brought his fleet into soundings on October 23rd, 1707, with a fresh gale at S.S.W., but hazy weather. In the evening he made more sail, believing that he saw the Lizard Light. Soon afterwards several ships made the signal of danger, and the *Association* struck on the rock called the " Bishop and Clerks." Two other ships, the *Eagle* and *Romney* were lost, with all on board.

Sir George Byng's flagship was saved through that officer's presence of mind when the rocks were almost under his main channels. This melancholy catastrophe created great consternation in England, for Sir Cloudesley Shovel was universally respected and was very popular.

Numbers of cases, equally melancholy, but of less notoriety, have occurred, and many others in which the danger has been imminent have scarcely reached the public ear. All had been referred to accident; and consequently, no attempt had been made to investigate their causes. Major Rennell came to the conclusion that they may all be imputed to a specific cause: namely, a current. The object of his paper was to investigate both the current and its effects, that seamen might be apprised of the times when they should expect to feel its influence; for at such times only it is likely to occasion mischief. The current that prevails at ordinary times is probably too weak to occasion a serious error in the reckoning, equal to the difference of parallel between the south part of the Scilly and the fair way up channel, such as a prudent commander, unsuspicious of a current, would choose to sail in.

As a midshipman on board the *Brilliant*, in 1757, Rennell sailed close along the coast of Spain; and he remembered, amongst the earliest hydrographical facts that were stored in his memory, that there was always a current setting round Cape Finisterre and Ortegal into the Bay of Biscay. The fact was confirmed to him in fuller detail by Captain Mendoza Rios, of the Spanish navy. This current sets to the eastward along the coast of Spain, and continues its course along the coast of France to the north and

K

north-west. Rennell supposed the original cause of the
current to be the prevalence of westerly winds in the
Atlantic. The stronger the wind, the more water
would be driven into the Bay of Biscay in a given
time ; and the longer the continuance of the wind, the
farther will the vein of current extend.

It is clearly proved that a current of water, after
running along a coast that suddenly changes its
direction, does not alter its course with the shore, but
the main body of it preserves for a considerable time
the general direction which it received from the coast
it last ran by. There is such a change in the direction of the coast on the French shore at Penmarch,
south of Brest; so that the current of the Bay of
Biscay continues its course N.W. by W. from the
coast of France to the westward of the Scilly Islands,
and to Ireland. In ordinary times its strength may
not be sufficient to preserve its line of direction across
the mouth of the English Channel. But that a
current prevails generally there can be little doubt,
and its degree of strength will be regulated by the
state of the winds. After a long interval of moderate
westerly winds it may be hardly perceptible; but after
hard and continued gales from the western quarter
the current will be felt in a considerable degree of
strength, not only off the Scillies, but also along the
south-west coast of Ireland.

Major Rennell was told by his friend Mr. Smeaton,
the builder of the Eddystone, that he tried an
experiment to ascertain the power of the wind on
water. In a canal four miles long the water was kept
up four inches higher at one end than at the other
by the mere action of a moderate breeze along the
canal. The effect of a strong north-west or south-

west wind on our own coast is to raise very high tides
in the English Channel, or in the Thames, and on the
eastern coasts, as those winds respectively blow,
because the water that is accumulated cannot escape
quickly enough by the Straits of Dover to allow of
the level being preserved. The Baltic, also, is kept
up two feet at least by a strong N.W. wind of any
continuance; and the Caspian is higher by several
feet at either end as a strong northerly or southerly
wind prevails. Therefore, as water pent up in a
situation from which it cannot escape acquires a
higher level, so in a place where it can escape the
same operation produces a current, and this current
will extend to a greater or less distance, according to
the force with which it is set in motion.

Major Rennell got his first idea of the existence of
this current in the Bay of Biscay when he was returning
from India on board the *Hector* (Captain Williams), in
1778. Between 42° and 49° N. they encountered very
strong westerly gales, but particularly between the
16th and 24th of January, when, at intervals, it blew
with uncommon violence. Rennell afterwards learnt
that the gale extended from Nova Scotia to the coast
of Spain. On the 30th of January the *Hector* arrived
within sixty or seventy leagues of the meridian of
Scilly, keeping between the parallels of 49° and 50° N.;
and about this time a current began to be felt, which
set the ship to the north of her intended latitude by
half a degree in two days, and thirty leagues to the
west. The wind being scant and light, they could not
overcome the tendency of this current. They were
not only sensible of the current by the observations
for latitude, but by ripplings on the surface of the
water and by the direction of the lead-line. The

K 2

consequence of all this was that they were driven to
the north of Scilly, and were barely able to shape a
course through the passage between those islands and
the Land's End.

The journal of the *Atlas*, East Indiaman (Captain
Cooper), furnished Major Rennell with still clearer
proofs both of the existence of the current and of
the rate of its motion, because Captain Cooper was
provided with chronometers, while the *Hector* was not.
The *Atlas* left the Isle of Wight on January 25th,
1787, and on the 27th was fifty-five leagues to the
westward of Ushant. On that day a violent gale of
wind began to blow from the south, changing
suddenly to west after eleven hours. The gale continued through the four following days. During this
long interval the ship was generally lying-to, with her
head N.W. The wind abated on the 1st of February,
still blowing from the S.W., and the ship's head was
kept N.W. The stormy weather returned the following day, and continued, with little intermission, until
the 11th. The weather then became more moderate,
and the ship proceeded on its way to the south.

It was found that by the 2nd of February the
Atlas had been set two whole degrees of longitude to
the westward of her reckoning since the 30th, at noon.
The *Atlas*, in fact, experienced a westerly current
from a point twenty-four leagues W.S.W. of Scilly to
four degrees of longitude W. of the meridian of Cape
Clear, where its influence was no longer perceptible.
There must also have been a northerly set, although it
is not recorded.

Major Rennell concluded his first paper with some
valuable practical sailing directions in crossing the
current. He advised the Government to send a

vessel, with chronometers on board, to examine and note the soundings between Scilly and Ushant. He believed that the existing chart of soundings was very bad : indeed, it could not be otherwise, considering the very imperfect state of marine surveying at the time when it was made. The surveyor so employed, he recommended, should note all the varieties of bottom, as well as the depths, the time of high and low water, the set of tides, and the currents. Such a survey, he concluded, skilfully conducted, might enable mariners to supply the want of observations for latitude and longitude, and, of course, to defy the current, as far as relates to its power of misleading them. These remarks of Major Rennell were frequently quoted by other writers, and they had a strong effect on the minds of the Lords of the Admiralty. It was, no doubt, due to them that the present excellent surveys of the Channel were undertaken.

In his second paper on the subject of this current, read on April 12th, 1815, Rennell submitted some additional evidence of great interest. He adduced the journals of several ships, the East Indiamen being all supplied with chronometers, showing that a current flows eastward along the north coast of Spain. Off Cape Ortegal, Admiral Knight found the current setting nearly along shore, at the rate of one mile per hour. It was undoubtedly this current which was the cause of the disaster to the *Apollo* in 1803. She sailed for the West Indies with a convoy of sixty-nine sail of merchant ships, and on April 2nd the *Apollo* struck the ground during the middle watch, to the utter astonishment of every one on board. When morning broke, it was seen that thirty

of the merchantmen were also on the rocks near
Mondego Bay, in Portugal. It was this disaster that
made Captain Markham, then one of the Lords of
the Admiralty, insist upon men-of-war being supplied
with chronometers. East Indiamen had been so
supplied several years previously.

The course of the current round the Bay of Biscay
had long been known to French navigators. One
circumstance, proving the northerly course of the
current, is very striking. The soundings in the Bay
of Biscay show little or no muddy bottom to the
southward of the River Gironde, but everywhere to
the northward. This shows that the mud of the
Gironde, Charente, and Loire is all carried northward
by a northerly current. The alluvial *embouchures* of
the rivers in general on this coast, and the positions
of the banks formed by them in the sea, also point to
the north and north-west as the effect of the same
current.

The most satisfactory proof, in the opinion of
Major Rennell, not only of the existence of a
northerly current athwart the mouths of the English
and Irish Channels, but also of its velocity in certain
periods, is a statement in a book, entitled "Joshua
Kelly's Treatise of Navigation," published in 1733. It
appears that in a voyage from the West Indies an
experienced commander observed that in about
48° 30' N., it became calm, and remained so for
forty-eight hours, with clear weather, so that he could
get good observations for latitude. In twenty-four
hours he found that his ship had drifted twenty miles
to the northward, and in the next twenty-four hours
he had drifted twenty-six miles in the same direction.
The current is the more dangerous because it is not

continuous, but only exists in strength at certain intervals.

The current, about the occasional force of which there is now no doubt, has received the name of Rennell's Current.

On June 22nd, 1809, Major Rennell read another paper before the Royal Society on the effect of westerly winds in raising the level of the English Channel, the subject having been suggested by the loss of the East Indiaman *Britannia* on the Goodwin Sands. He attributed the loss to a current produced by strong westerly winds in raising the level of the English Channel, and the escape of the superincumbent waters through the Straits of Dover at a time when the Northern Sea was at a lower level.

In those days Messrs. Laurie and Whittle were the principal publishers of charts and sailing directions, and they took care that Rennell's paper on the Bay of Biscay current should be made known to seamen by being published and widely circulated. It was this firm which presented Major Rennell with " Kelly's Book of Navigation"; and in acknowledging his receipt of the present, in a letter dated January 20th, 1809, he said :—" Having been accustomed to navigation books since the year 1756, I must have seen twenty or thirty different ones, but I never met with Kelly's work. I have now perused it, and derived much valuable information from it. Accept my best thanks for the honour you have done me in putting forth my current paper to so much advantage. There is nothing that gratifies me more than the testimony of being useful when that testimony comes from intelligent and useful persons."

Although Major Rennell was unable to complete

his system by reviewing the currents of the Pacific
and other oceans, he had studied the whole subject
with great care. We find evidence of this in his
private correspondence; especially in letters to naval
friends of a younger generation, such as Captain
Smyth and Captain Francis Beaufort, the eminent
surveyors.

In a letter to Captain Beaufort, dated June 11th,
1824, Rennell wrote:—" I return Captain Basil Hall's
book, with many thanks for the use of it. I have not
read anything with more satisfaction. It is written
with much taste and judgment, and is full of character.
I have compared the place where he saw ice off Cape
Horn with other reports, and I am confirmed in my
opinion that the ice there is brought by a S.W.
current from the ice latitude, passing between South
Shetland and Cape Horn. The immense ice island
seen by Captain Hall shakes our received system of
forming icebergs, unless there are lands towards the
Pole. The general currents in the South Pacific
resemble a good deal those in the South Atlantic
and the Equatorial Atlantic. A current along the
shore of Chili and Peru is caused by the prevalent
south wind. Then the S.E. trade raises *its* current,
which shoots off to the N.W., conforming to the line
of the coast, and growing stronger so as to become
rapid at the Galapagos. Captain Basil Hall must be
an able man. He is perhaps educating for future
great purposes."

Rennell had a European reputation as an authority
on winds and currents, and when Baron Humboldt
left Paris for the purpose of settling finally in Berlin,
he took London on his way, on purpose to have the
advantage of conversations with the great hydro-

grapher. In a very interesting letter from the late
Sir Edward Sabine to Major Rennell, dated April 12th,
1827, there is the following passage :—" I hope that
M. de Humboldt, who leaves Paris this day for
London, will not be disappointed in the expectation I
have given him that he will find you in tolerable
health. One of the two principal objects which have
induced him to take London in his way to Berlin is
to converse with you on the subject of currents and
temperatures of the sea, on which he has been latterly
thinking and seeking out facts even more diligently
than formerly."

When Major Rennell had completed his wind and
current charts, steps were taken for their publication.
They consisted: I., Of the Eastern Division of the
Atlantic ; II., The Western Division of the Atlantic ;
III., Southern Africa, with the Lagullas Current ; IV.,
Currents between the Indian and South Atlantic
Oceans ; V., The Gulf Stream ; VI., A General Index
Chart ; VII., Rennell's Current. In 1827, his staunch
friend, Lord Spencer, offered to represent the importance of publishing these charts to the Duke of
Clarence, who was then Lord High Admiral. The
offer was accepted, and Lord Spencer spoke strongly
on the subject, his opinion being supported by Mr.
John Barrow, the Secretary of the Admiralty ; but it
was left to Major Rennell's daughter, Lady Rodd, to
carry out the intentions of her revered father after
his death. " If," she wrote in her preface, " the prodigious mass of information accumulated in the charts,
with its skilful exposition in the memoir, should lead
to the prevention of shipwreck, and augment the
resources of seamen, the noble purpose of my father's
exertions will have been most amply attained." She

dedicated the work to King William IV., and mentioned that she had received his Majesty's commands to state that when Lord High Admiral his attention was first drawn to the great scope and importance of the undertaking through the discerning zeal of Earl Spencer. The editorial work was entrusted by Lady Rodd to Mr. John Purdy.

Most of his hydrographical work was executed by Major Rennell during the last twenty years of his life ; but he had been making collections and had taken a deep interest in the subject since he was a midshipman. His charts and memoir were invaluable at the time, and he was undoubtedly the founder of the science of oceanography, or, as it was called by Maury, "the physical geography of the sea."

Major Rennell was offered the post of first Hydrographer to the Admiralty, but he declined on the ground that he valued his literary pursuits too much to accept any official appointment which would take him from them. The post was then given to Dalrymple.

CHAPTER IX.

Ruins of Babylon.—Identity of Jerash.—Shipwreck of St. Paul.—Landing of Cæsar.

WHEN Major Rennell had passed his sixtieth year, he was still full of energy, and capable of long and sustained literary effort, although his constitution had never recovered from the effects of wounds and fevers from which he suffered in India. A relation of his wife thus described him at this time:—" He possessed the distinguishing mark of a great man—simplicity. He had the faculty of disguising and making others forget his own superiority, and he was thus spoken of by a contemporary: 'In his intercourse with his friends he possesses a remarkable flow of spirits, and abounds with interesting subjects of conversation; at the same time, as to whatever relates to himself, he is one of the most diffident, unassuming men in the world. His delicate constitution and his attachment to domestic life caused him to retire a good deal from general society, but he received his friends every forenoon at his own house.'

" Major Rennell was of middle height, well proportioned, with a grave, yet sweet, expression on his countenance, which is said to have conciliated the regard of all he spoke with. The miniature by Scott, painted for Lord Spencer, at about the age of sixty, represents him sitting in his chair with folded arms, as in reflection. The forehead is of remarkable height, shaded by abundance of grey hair; the nose long and finely cut, the eye mild and sunk below a massive brow." *

He lost his wife after a happy union of nearly forty years. She died at Brighton on the 2nd of January, 1810, and was buried at Harrow. She was described by one of her nieces as a quiet and most amiable woman, thoroughly domestic. She used to spend one day in every week at the Middlesex Hospital, which was close to her house in Suffolk Street. Their eldest son appears to have been a man without energy or ambition, and of very retiring habits. He lived near Enfield, where he died on the 25th of December, 1846. The second son, William, who was in the Bengal Civil Service, died at Futtyghur in July, 1819. He was married, but left no children, and his wife followed him to the grave within a few months.

From that time all Major Rennell's affections were concentrated on his daughter Jane and her children. She was beautiful and refined, possessed of great talent, and was always present at her father's morning *ré-unions* of learned men. On October 5th, 1809, Jane Rennell was married to Captain John Tremayne Rodd, R.N. The acquaintance arose through the intimacy of Major Rennell with Captain Wallis, who made the famous voyage round the world in 1766-68. Wallis was appointed a Commissioner of the Navy Office in 1780, and took a house in Seymour Street. He had married Miss Hearle, of Penryn, whose sister

* " Memorials of the Thackeray Family," by Mrs. Bayne.

Jane was the wife of Francis Rodd, of Trebartha, in
Cornwall, and mother of John Tremayne Rodd.
Commissioner Wallis had charge of young Rodd's
education, and put him into the navy. He became
a captain in 1788, and it was at the house in Seymour
Street that he first met Miss Rennell. Their honey-
moon was passed at Milton Bryan, the seat of her
father's old friend Sir Hugh Inglis, in Bedfordshire.
Soon afterwards they took 45, Devonshire Street,
Portland Place, to be near Major Rennell, and here
their children were born—James Rennell Rodd in
1812, Alicia, Frances, and Wilhelmina. Their father
was created C.B. in 1815, and became Rear-Admiral
Sir John Tremayne Rodd, K.C.B., in 1825. In about
1820 the Rodds removed from Devonshire Street to
No. 40, Wimpole Street. During the last years of his
life Major Rennell's affections were centred on his
beloved daughter and her children, from whom he
received the most constant and assiduous attention,
which brightened his declining years and alleviated
the sufferings of old age, aggravated by the tortures
of the gout.

Besides his great labours connected with hydrography and the geography of India, Western Asia
and Africa, Mr. Rennell communicated several papers
to the Society of Antiquaries, which were printed in
the *Archæologia*. The earliest of these was on the
topography of ancient Babylon, suggested by the
observations and discoveries of John Claudius Rich.
It was a subject which had previously engaged much of
Rennell's attention while at work on the "Geography of
Herodotus," and it appeared to him that some of
the inferences of Mr. Rich, who was British Resident
at Bagdad, were contrary to the evidence of ancient

writers. According to the uniform consent of antiquity, Babylon was of vast size, and was built on
both sides of the River Euphrates, which flowed
through the centre of it. Herodotus mentions the
immense embankments built to confine the Euphrates
in its proper channel. The city was in form an exact
square, each side being a hundred *stadia* long, with
ditch and a wall two hundred cubits high. A
hundred brazen gates pierced the walls, and the city
was divided into two distinct parts by the river. In
each of the two quarters of the city there was a vast
edifice: the royal palace in one and the temple of
Belus in the other. Diodorus further mentions a
bridge across the Euphrates and two palaces at each
end of the bridge.

The ruins of Babylon commence eight miles north
of the town of Hillah, and a little south of the village
of Mohawill. They were examined in detail by Mr.
Rich in 1811, and by Sir Robert K. Porter in 1818;
and had previously been visited by Pietro della Valle,
Beauchamp, and Niebuhr. Mr. Rich reported that
all the ruins of the great city were on the east side of
the river, consisting of three principal mounds, in a
north and south line, called *Mujallibah*, *Kasr*, and
Amran ibn Ali, from a small mosque on the top.
At a distance of eight miles to the south-west is the
vast mound called *Birs-i-Nimrûd*. The two mounds
called *Mujallibah* and *Birs-i-Nimrûd* are about the
same size, but the last is a hundred feet higher.
Rich discussed the question whether the *Mujallibah*
or the *Birs-i-Nimrûd* was the great temple of Belus,
it being nowhere stated by any ancient writer in
what quarter of the city that temple stood. Rich
decided in favour of Birs-i-Nimrûd.

Major Rennell, in the paper discussing Mr. Rich's report, pointed out that if, as Mr. Rich stated, all the ruins are on one side of the Euphrates, either the river must have altered its course or else the statements of all the ancient writers must be wrong. It appeared to Rennell that, in considering the sites and distribution of the ruins of this ancient city, two circumstances arise that cannot be lost sight of. One is the probable change in the course of the river, and the other that the whole of the mounds in the neighbourhood do not belong to ancient Babylon. Without a subsequent change in the course of the Euphrates, the descriptions of Herodotus cannot be reconciled with the position of the ruins. All the ruins are now on the eastern side, although the principal palace and the hanging gardens certainly stood on the opposite side to the tower and temple of Belus.

Mr. Rich states that beyond Mohawill the ruins of Babylon commence, the whole country between that village and Hillah having traces of buildings, Hillah being nine miles from Mohawill. The Mujallibah mound is five miles from Hillah, and about four from Mohawill, so that it is nearly in the centre of the space occupied by heaps of ruins. Mr. Rich mentioned a large canal at Mohawill, which Rennell supposes to have been the exterior ditch of the city; and the mound of Mujallibah would be near the centre, where the tower and temple of Belus are described to have stood by Herodotus. The principal mass was probably formed of sun-dried bricks, with a coating of those baked in the furnace. The latter have been taken away for use in modern buildings; the former have been dissolved by rain, and carried away by the winds in a pulverized state. The tower of Belus was in ruins in the time of Alexander the Great. Herodotus places the temple in the central part of the quarter on one side of the Euphrates, and the great palace in the same position on the other side of the river. On the supposition that the river has altered its course, ruins exist which might be those of the palace. On Mr Rich's plan there are four mounds, besides Mujallibah, three of which are described in his memoir; and of these, the one named *Kasr* is supposed by him to be the palace: a view in which Major Rennell fully concurs. The opinion is founded on the vast area it covers, the solidity of the walls, the superior quality of the workmanship, and the articles that have been found there.

Kasr is situated a good mile from *Mujallibah;* but in order to make the positions agree with the descriptions of Herodotus and Diodorus, a change of course of the Euphrates must necessarily be admitted. Nothing is more common than such changes during a course of many centuries, in an alluvial soil, especially when the river is subject to be choked by the falling of ruins. Major Rennell observed that in the course of his observations he had found that no cause operates more powerfully in diverting the courses of streams than fallen ruins. Admitting this possible change, Rennell suggests that the mound to the north-east of Kasr is the lesser palace, which was separated from the other by the river and a bridge. The ancient bed can be traced through a course of about a mile and a half, with a width of 150 yards, which is the same width as that of the present river. "Few persons conversant with the nature of rivers and their changes will for a moment hesitate to receive the internal evidence, contained in Mr. Rich's plan and description, in proof of the ravine there shown being a deserted bed of the Euphrates, and proving the tendency of that river, in common with all others that pass through alluvial soils, to vary their courses, unless art be employed to prevent it."

Major Rennell then discusses the questions relating to the Birs Nimrûd mound, on the western side of the Euphrates. He does not think that this mound has the character of an artificial work, nor that the masonry on its summit is part of such a structure as the tower of Belus is described to have been. He rather looks upon it as a building which covered the summit of a conical hill. But whether it be natural or artificial, the Birs Nimrûd is so far distant from the centre of the ruins of ancient Babylon that it cannot have been the temple and tower of Belus. Birs Nimrûd is seven and a half miles from the Kasr mound, and eight and a half from Mujallibah. It cannot even be included within the area of Babylon.

Mr. Rich replied to Major Rennell's paper in 1817, in a rejoinder, in which he entered more fully into the details collected during his examination of the ruins. It has since been discovered that Birs Nimrûd represents the site of the ancient Borsippus outside Babylon; and Sir Henry Rawlinson says that the name Borsippa is found upon the records of the obelisk from Birs-Nimrûd.

It is astonishing how Major Rennell took in all the complicated details of plans such as that of the ruins of Babylon; with what acuteness and sagacity he compared all the statements of ancient and modern writers; how completely he realised the positions to himself; as if he had personally surveyed them, and how he built up the whole in his mind's eye with

L

wonderful accuracy. This faculty must have given the highest kind of enjoyment to his studies, and have imparted pleasure which can only be enjoyed by those who combine great powers of imagination with the acquired habit of critical examination, guided by the thoughtful study of a mind always well under control.

Of course, Rennell could only base his conclusions on the materials within his reach. Knowing the passages of Herodotus, Diodorus, and Strabo by heart, he compared them with the more recent narratives of Pietro della Valle, Niebuhr, Beauchamp, Rich, Ker Porter, and Buckingham, and with the plan drawn by Rich, who examined the mounds and their positions with great care. He had very fully discussed the various questions relating to the ruins of Babylon in his work on Herodotus, and the deep interest he took in them led to his communicating the paper on Rich's work to the Society of Antiquaries.

The final conclusions on these questions, as the results of later research, are all in favour of Major Rennell's arguments. The *Birs-Nimrûd* was held by Rich, Ker Porter, and others to be the tower of Belus, contrary to the views of Rennell. From the inscriptions discovered on the spot, and from other documents of the time of Nebuchadnezzar, there is good reason for the belief that it is the site of Borsippa, and entirely beyond the limits of Babylon. Rennell's opinion that the Euphrates in ancient times flowed in a different bed, so as to have the mounds marking the ruins on either bank—as described by ancient writers—has also been confirmed by later research. Canon Rawlinson holds that this is the only way to reconcile the ruins as they exist with the descriptions of ancient writers. A large canal, called Shelil, intervened between the *Kasr* mound and the *Mujallibah* of Rich, also called *Babil*. The canal may easily have been confounded by Herodotus with the main stream of the Euphrates. This would have placed the two principal buildings, the palace and the temple of Belus, on opposite sides of the canal. Canon Rawlinson, therefore, agrees with Major Rennell in identifying the *Babil*, or *Mujallibah* mound, with the great temple and tower of Belus. He further agrees with Rennell that the *Kasr* mound is the great palace of Nebuchadnezzar. Layard says that the ruins contain traces of architectural ornament, such as piers, buttresses, and pilasters.

This is a very striking example of the accuracy of Major Rennell's judgment. It is far from uncommon to find an explorer or a student seizing upon some point as a hobby, and then striving to force all the other evidence into harmony with it. The way in which Rich and Ker Porter argued that Birs-Nimrûd was the temple of Belus is a case in point. Nothing could bring it to the centre, or anywhere near the centre, of Babylon, or even within the walls of Babylon, without the most extravagant stretching of the facts to fit into a theory supported by no evidence. The method of Rennell forms a strong contrast. Having all the evidence in his mind, he carefully studied the plan of the ruins, and harmonised the two classes of *data* with such unerring judgment that his views have been confirmed by more recent research.

Another question which interested Major Rennell was the identity of Jerash; and he read a paper at the Society of Antiquaries on June 17th, 1824, "concerning the identity of the architectural remains at

L 2

Jerash, and whether they are those of Gerasa or
Pella." The magnificent Grecian ruins at Jerash, in
the country beyond the Jordan, were discovered by
Dr. Seetzan in 1806, and afterwards described by
Burckhardt. Captains Irby and Mangles also executed
a careful survey. The ruins at Jerash are those of a
fine city in a beautiful situation. They are on two
sides of a valley with a stream running through it,
and consist of two main streets crossing each other at
right angles. The streets were lined with double
rows of pillars, some of which are Ionic and some
Corinthian. There are remains of pavement with the
marks of chariot wheels, and elevated ways on each
side for foot-passengers. There are two theatres, two
grand temples—one dedicated to the Sun—five or six
smaller temples, and an Ionic oval space three hundred
and nine feet long. Upwards of two hundred and
thirty columns are still standing, and there is a very
large reservoir for water at a distance of two hundred
yards. The inscriptions, which are numerous, are
chiefly of the date of the Emperor Antoninus Pius.
The ruins at Jerash are considered to be quite equal
in beauty and interest to those of Palmyra.

Notices of Pella and Gerasa are chiefly derived
from the works of Josephus and one or two early
Christian writers. Josephus describes the position
and extent of the region called Peræa, which included
Pella and Gerasa. Pella must have been near the
northern extremity of Peræa, a few miles from Jabesh-
Gilead, and not far from the southern shore of the
Lake of Tiberias. But the ruins of Jerash are equi-
distant from that lake and the Dead Sea, or near
the central part of the eastern border of Peræa. It
is, therefore, clear that the ruins at Jerash cannot be

those of Pella. Major Rennell then proceeds to consider how far the ruins may be identified with the ancient Gerasa. Josephus mentions Gerasa as being on the eastern border of Peræa; and Eusebius says that the Jabbok River flowed between Philadelphia and Gerasa, at a distance of four miles from the latter. On the whole, Jerash may be allowed to occupy the general position assigned by Josephus and Eusebius to Gerasa.

Rennell considered that very great misconceptions had existed among modern geographers respecting the situation of Gerasa, until the discovery made by Dr. Seetzan, in 1806. Rennell is inclined to identify Jerash with the ancient Gerasa: a conclusion which is now universally accepted. The place must have risen into importance in the times of the Antonines, after the great descriptive geographers of antiquity had ceased to write; so that history tells us nothing of the prosperous period of Gerasa, when its stately temples and theatres were erected. It is evident that its importance was due to its having been an emporium of commerce, through which the great trade route passed which was afterwards diverted to Palmyra.

Major Rennell suggests that the confusion respecting Gerasa may have arisen from the occurrence of the word Gergesenes, in the Gospel of St. Matthew, originating the idea that Gerasa was situated near the Lake of Tiberias; but as both St. Mark and St. Luke relate the same events to have taken place among the Gadarenes, it may be assumed that the word Gergesenes is an error of a copyist. There can be scarcely a doubt that the extensive ruins at Am-Keis, on the southern coast of the Lake of Tiberias, are those of the city and fortress of Gadara, mentioned by

Josephus as the ancient capital of Peræa. The
connection of Gadara with sacred history is also
interesting. It contained inhabited tombs, as in the
time of the visit of Jesus to this neighbourhood.
Captains Irby and Mangles, who were there in 1818,
have given an account of these singular dwellings,
from which the dead must have been expelled eighteen
hundred years ago to accommodate the living. Those
travellers found no other inhabitants of Gadara but
the dwellers in ancient tombs. These tombs were
excavated out of the living rock, near the top of the
mountain, and one of them, in which the travellers
were hospitably received by the sheikh, was capacious
enough to contain his family and cattle, as well as his
guests. The sepulchres appear to be very numerous.

Rennell points out that Gadara and its sepulchres
are seven miles from the shores of the lake, but that
our Saviour is not said to have visited the city of the
Gadarenes. He merely landed within its territory
on a shore of the lake over against Galilee, which
agrees with the territory of Gadara. He came in a
vessel from the side of Capernaum, and was accosted
by the man from the tombs immediately on landing;
but the tombs from which the man came were in a
city among the mountains. St. Luke says:—"When
Jesus went forth to land there met him out of the city
a certain man who had no abode in any house, but in
the tombs"; and St. Mark—a man "who had his
dwelling among the tombs, and always, night and day,
he was in the mountains and in the tombs." From
these passages we gather that the man from the
tombs came from a city which was situated in the
mountains; and the particulars seem to point clearly
to Gadara.

The question whether the island on which St. Paul was wrecked was Malta or an island in the Adriatic, and numerous collateral points connected with it, had very special interest for Major Rennell. It combined hydrography with antiquarian research, and, therefore, aroused in him the spirit of enquiry, both as a sailor and a comparative geographer. In January, 1824, he read a paper before the Society of Antiquaries, "On the Voyage and Place of Shipwreck of St. Paul, in A.D. 62." From the earliest times the island of Malta had been referred to as the place of shipwreck; and this was very satisfactorily demonstrated by Bochart, in his "Chanean," written in about the middle of the seventeenth century. The person who first disputed it was Ignacio Giorgi, a Benedictine monk, of the Island of Meleda, on the coast of Dalmatia. In 1730 he wrote a dissertation to prove that the shipwreck of St. Paul happened at this Dalmatian island, and not at Malta. The argument turns almost entirely on the application of the term "Adria" to the present Adriatic Sea alone. Forty years afterwards the learned Jacob Bryant took up the same view. Bryant was an antiquary of some note in his day, and the author of a curious work, entitled "Analysis of Ancient Mythology"—a work denying the existence of Troy—and other eccentric productions. He was born in 1715, and lived to the age of ninety. His argument respecting St. Paul was that the ancient historians and geographers down to the time of the elder Pliny confine the term "Adria" to the present Adriatic. Rennell admits that this is true, but he adds that Ptolemy, only seventy years after Pliny, extends the name "Adria" to part of the Mediterranean, and mentions it repeatedly.

Major Rennell, in considering the question, weighed the evidence of the different authorities with great care. He regarded the circumstances of the navigation, as related by St. Paul, in respect to the position of Melita, as perfectly conclusive in themselves; but as the name of "Adria" was the stumbling-block, he set forth, in the first place, the authorities for the position of that sea. Ptolemy distinguishes the *Gulf* from the *Sea* of Adria. The former bordered on the eastern side of Italy, while the latter was to the south, forming the north part of the middle basin of the Mediterranean Sea. The extent of the Sea of Adria southwards, beyond the gulf of the same name, is fixed by four passages in Ptolemy. In one place he says: "Italy terminates towards the south, on the shore of the Sea of Adria"; in another: "Sicily is bounded on the east by the Sea of Adria"; and again: "The Peloponnesus is bounded on the west and south by the Sea of Adria"; and, finally: "Crete is bounded on the west by the Sea of Adria." Ptolemy wrote seventy or eighty years after the voyage of St. Paul, when the term Adria was certainly applied to the sea between Sicily and Crete. Although Malta is about sixty miles to the south of a line drawn from Sicily to Crete, yet the boundary of a sea cannot have been so strictly defined, and the ship of St. Paul must have been within the strict limits of Adria during part of the storm.

But, apart from the position of the Sea of Adria, ·Major Rennell argued that the whole narrative of the author of the Acts—considered with reference to geography, the winds, and other circumstances— affords conviction that Malta was the place of shipwreck. It was at some distance short of the Island of Clauda (the Gozza of Candia) that the ship was assailed by the tempestuous wind called "Euroclydon," which is the Levanter of modern times; and here the disputed part of the voyage commences. Major Rennell corresponded with Captain Francis Beaufort on the subject of the Levanters, and in June, 1823, the eminent naval surveyor wrote a luminous paper on the subject for the use of his illustrious friend. The Euroclydon, they concluded, was a strong easterly wind, and that off the south coast of Crete, coming off the land, it had some northing in it.

The undergirding of the ship was the frapping with hawsers, which was resorted to when the frame was weakened by decay or accident, and there appeared to be danger of its opening. Thus strengthened, the ship was made to run before the wind under very reduced sail. Her course would, therefore, be south of west, and would lead towards the Greater Syrtis and its dreaded quicksands. So to avoid that danger they took in all sail, with the object of lessening their rate of motion. This appeared to Major Rennell to be the true explanation of the fears that are mentioned respecting the quicksands. The surveys of Admiral Smyth, when he commanded the *Adventure*, showed that the shores of the Greater Syrtis were quite changed in their character since ancient times. A firm sandy soil has replaced the old quicksands: a change to be accounted for by the operation of the surge of the sea in northerly and north-westerly storms during many centuries. They have thrown up and spread the sand over the lands contiguous to the margin so as to raise the surface beyond the reach of the ordinary level of the sea, thus preventing it from being dissolved or melted into quicksand, as

formerly. The different state of the Goodwin Sands
at high and low water respectively affords a practical
illustration of this. At low water it is so firm as to
be with difficulty penetrated with a pointed piece of
wood, while towards high water Mr. Smeaton reported
that it would not bear the weight of a man.

Fourteen days elapsed between the commencement
of the storm and the arrival of St. Paul's ship at
Melita, when they were "driven up and down in
Adria." The narrative does not give the direction of
the wind after the time when the Euroclydon is first
mentioned ; but as the Levanters are said sometimes
to continue for a fortnight, it appears most probable
that the storm generally blew from the eastward, in
which case Malta would certainly be the Melita of
St. Paul's shipwreck, as that island lies almost west
from Crete. It is quite certain that nothing but a
long-continued storm from the southward could have
driven the ship, in the specified time, three hundred
and fifty to four hundred miles to the northward,
so as to reach the Dalmatian Melita, from her course
between Crete and the Straits of Messina; nor could
this have taken place without land being sighted.

They found in the port of Melita a ship of
Alexandria, called the *Castor and Pollux*, bound for
Rome, on board of which St. Paul was afterwards
embarked. She is said to have wintered at Melita:
that is to say, she had taken refuge there to avoid
being at sea during the autumnal equinox. Much
light is thrown on the subject of wintering by the
proceedings on board the ship of St. Paul at this
very season. That ship, it appears, was on its way to
the port of Phenice, in Crete, in order to winter, and
must have been near it when caught by the storm

which the mariners had expected and were endeavour-
ing to avoid. They had previously suffered much delay
from light and contrary winds, and the season was
far advanced. Hence, Major Rennell concludes that
as both ships were acting on the same system, it was
the ordinary practice of the time. On the approach
of the expected stormy weather, they sought shelter in
the nearest port. On this principle, the ship of St.
Paul had first entered the port of the Fair Havens in
Crete, but quitted it to go to Phenice as a more secure
anchorage, contrary to the advice of St. Paul.

Applying this practice to the *Castor and Pollux*,
Major Rennell furnishes a clinching argument against
the Melita of Dalmatia. She was making the voyage
from Alexandria to Rome. To suppose that a ship so
circumstanced should have proceeded to the islands
off Dalmatia for the purpose of obtaining shelter
during the season of the equinox involves no small
degree of absurdity; for the Dalmatian Melita is
little less distant from a ship's track between Crete
and the Straits of Messina than Rome itself, which
was the final destination of the ship; so that she
must have gone nearly as far out of her way in order
to obtain casual shelter as she had to go to Rome, the
place of her final destination. Rennell held that this
circumstance alone was decisive against the claim of
the Dalmatian Melita. His conclusion was that the
evidence on the side of Malta far outweighed that on
the other side, even if the argument rested on the
circumstances of the narrative alone, without any
regard to the meaning or the application of the term
"Adria."

On the 4th of May, 1826, Major Rennell read a
paper before the Society of Antiquaries " Concerning

the Place where Julius Cæsar landed in Britain."
The first military expedition of Cæsar to Britain took
place in the autumn of 55 B.C., his place of embarkation in Gaul being Portus Itius for his infantry, and
another port eight miles distant from it for his cavalry.
D'Anville decided that the Portus Itius was Wissant
Bay, between the Capes Grisnez and Blancnez, directly
facing Dover. Cæsar gives no intimation whether he
landed on the eastern or southern coast of Kent, and
opinions have differed widely. D'Anville, followed by
Beale, Porter, and Lewin, held that the landing was at
Romney Marsh, while Rennell and others maintain
that Deal Beach was the place of disembarkation.
This was a question that would have a special interest
for Major Rennell as a hydrographer. He had studied
the tides and currents of the British Channel during
many years with close attention, and such a problem
as the place of landing of Cæsar would have had a
peculiar fascination for him. The wind was favourable, but as a southerly wind would be fair either for
Deal or Romney, this proves nothing, and the enquirer
must turn his attention to other circumstances.

Cæsar first sent Caius Volusenus in a ship to
reconnoitre the opposite coast, and discover all that
was possible without disembarking. He was absent
five days. On his return, Cæsar set sail at about midnight of the 24th of August, with a south-westerly
wind. He arrived off Dover at ten in the forenoon of
the 25th, and saw the enemy posted on all the hills.
He then enquired of Volusenus how far the cliffs
extended, and he ascertained that a few miles further
along the coast there was an open and level shore.
Now this is the case on either side of Dover, whether
the landing was at Deal Beach or at Romney Marsh.

Cæsar waited until about three in the afternoon, when the tide and wind were both in his favour, and then gave the signal to weigh anchor. It will, therefore, be seen that the question depends on the direction in which the tide was running at three in the afternoon on the 25th of August. The distance from Dover to the flat coast where the landing was effected is stated to have been eight Roman miles, which agrees with the distance between Dover and Walmer Castle, reckoned along the coast; but on the west side such an open coast as an army would be landed on is twelve or thirteen Roman miles from Dover.

The landings on Cæsar's first and second expeditions are stated to have been at the same place ; and in the account of the second expedition a fact occurred which alone, in Major Rennell's opinion, proves that Cæsar landed to the eastward of Dover. It is stated that in this second expedition the fleet left Gaul at sunset, perhaps after seven, as it was in the summer. At midnight the wind, which had been a gentle breeze from the south, fell, and the vessels were left to the resource of their oars. In the morning it was found that they had been carried by the tide far beyond their intended point, being the landing-place of the preceding year, and they reached it in the forenoon of that day by availing themselves of the returning tide. The fact to which Major Rennell alludes is this: It is said that on the morning after their leaving Gaul they saw the land of Britain on their left hand. Now, this left hand must be spoken of in reference to the general direction of their course from Gaul towards their former landing-place in Britain. In that case the land in question could be no other than the eastern side of Kent, seen on the left of the voyagers at daybreak ; consequently, Cæsar must have been at that time on the eastern side of Dover.

Thus, although the account of the first expedition taken alone affords no positive evidence concerning the particular side of Dover on which Cæsar landed, yet, taken with the details of the second expedition, and with the fact that both landings were made at the same place, they show that Cæsar landed on the eastern side of Dover, and fix the place at Deal Beach. Of course, owing to changes on the coast, the margin of the ancient beach on which Cæsar landed must now be very far inland, as well as considerably raised.

This original argument of Major Rennell, which really settles the question, is one other instance of his sagacity and critical insight. The controversy has been continued to the present day; but Rennell's argument—used by others without any acknowledgment—has proved too strong for the advocates of Romney Marsh, and it is now very generally admitted that Deal was the true place of Cæsar's landing. Mr. Vine has taken much pains to establish the accuracy of this conclusion by a careful study of Cæsar's line of march from Deal, which he traces by comparing the descriptions of Cæsar and the physical aspects of the country with much learning and ingenuity.

The direction of the tide at three o'clock in the afternoon, when Cæsar was at anchor off Dover, would be equally conclusive; but on this point there will always be doubt. Halley calculated that the tide would begin to flow to the east at about half-past three on the 25th of August, B.C. 55. He first proved that this was the day on which Cæsar landed, the year being the Consulate of Pompey and Crassus,

A.U.C. 699; the month is fixed by the description of the season, and the day by the phase of the moon. Sir George Airy, however, put forward an argument which, if correct, is fatal to Deal as the landing place. He doubts whether Cæsar knew the exact date of the full moon; and as the highest tide takes place a day and a half after the full moon, Cæsar must have made a mistake in placing the two phenomena at the same moment, either as regards the day of the full moon or that of the highest tide. So he concludes that the landing must have taken place on the second, third, or fourth day before the full moon. But this may be admitted without affecting the calculation of Halley. The question is to determine the day at the end of August when the tide changed at Dover at half-past three in the afternoon. This would be on the 25th of August, when Cæsar's fleet would have been carried forward by the current of the rising tide.

Professor Burrows has remarked that the changes along the coast cannot fail to vitiate the calculations which have been held to decide the place of Cæsar's landing in the Cinque Ports districts. The depth of the channel may have largely varied, while the space over which the tides travel must be at least two miles wider than it was some two thousand years ago. Consequently, the point of meeting of the north and south tide streams cannot possibly be exactly the same; yet this is the assumption under which all these calculations have been made. On the whole, however, the evidence of the tides, so far as it goes, is in favour of the landing at Deal.

But it is Major Rennell's argument, from the circumstance related in the second expedition, which really settles the question, quite independently of any

calculation relating to the tide at Dover in the afternoon of the 25th of August. From the position
whence at daybreak Cæsar saw Britain on the left
hand, in the second expedition, after having been
carried north and east by the tide, his fleet, with hard
work, could have reached Deal by noon; and it could
not possibly have reached Hythe and Romney Marsh
by noon.

Major Rennell's apparently desultory studies, which
led to his writing the papers which have been published in the *Archæologia*, all arose out of important
work on which he had been engaged. The discussions
on the ruins of Babylon and at Jerash are connected
with his work on "Herodotus," and were the natural
consequences of the receipt of fresher information.
The disquisitions on the shipwreck of St. Paul and
the landing of Cæsar were partly the outcome of his
hydrographical researches, combined with a love for
the solution of questions where history is dependent
on geography for its due comprehension. Of the
numerous services done by Major Rennell for geography and hydrography, one of the greatest is the
method he introduced of making his science essential
to the study of history. When his data were accurate—
and he generally succeeded in making them so—his
method being logical and correct, his conclusions were
right in almost every instance.

CHAPTER X.

SIR JOSEPH BANKS, with his patronage and his hospitable *réunions* in Soho Square, and Major Rennell, with his works and investigations, his advice and assistance ever ready, and his forenoon receptions in Suffolk Street, practically formed, with their friends, a working Geographical Society during the fifty years which preceded its actual inauguration. After the death of Sir Joseph, in 1820, Major Rennell was the acknowledged head of British geographers for the next ten years. Travellers and explorers came to him with their rough work, projects were submitted for his opinion, reports were sent to him from all parts of the world. He presided over the labours of geographers, formed a central rendezvous for help and advice; and on his death, the formation of a Geographical Society to supply his place became a necessity.

There was nothing, in the days of Banks and Rennell, which did more to advance the general interests of geography than the numerous opportunities of intercommunication which were offered by the hospitalities of Soho Square and Suffolk Street, where travellers and students exchanged ideas, and became known to each other. When Sir Joseph had passed away, and Major Rennell had become too old and

M

infirm to take any lead at such meetings, the want became generally felt. The machinery would grow rusty. It was still a necessity "to lubricate the wheels of science." The originator of that well-known and frequently used figure of speech is uncertain. No one, in after years, ever performed that function more gracefully and more efficiently than the accomplished nobleman who presided over the Geographical Society just a quarter of a century after the death of Rennell. The Earl of Ellesmere held that opportunities of occasional intercourse, at dinners or evening parties, were so important that he requested, after he ceased to be president, to be allowed to retain the honour, the privilege, and the singular pleasure to himself of continuing to promote such intercourse. He then said that "to lubricate the wheels of science" was an expression of Lord Stowell. No doubt it was; still, we must go farther back for the originator of that happy phrase.

Lord Stowell was almost an exact contemporary of Major Rennell, being three years younger, and sur-viving him for six years; but in 1826 old age pre-vented the latter from continuing the duties which he had performed so well and for so many years. Then it was that the geographers and travellers sought for a remedy to supply the want they soon began to feel. The first to give expression to this feeling was Sir Arthur de Capell Broke, a traveller and an author, well known for his adventures in Sweden and Norway, in Spain and in Morocco. He conceived the idea of forming an agreeable dining club, composed entirely of travellers. The world was to be mapped out into so many divisions, corresponding with the number of members, each division being represented by at least

one member, as far as it might be practicable, so that
the club collectively should have visited nearly every
part of the known globe. He first communicated
his idea to one of Rennell's most intimate friends,
Colonel Martin Leake, and then to Captain Mangles,
R.N., to "Legh, the traveller," who had been up the
Nile beyond the cataracts, and to Lieut. Holman, R.N.,
the well-known blind traveller. They prepared a list
of distinguished men, and a circular was sent to these
personages, signed by Sir Arthur Broke. The principal object of the meetings was announced to be
the attainment at a moderate expense, of an agreeable
friendly and rational Society, formed by persons who
had visited all parts of the world. The club received
the name of the Raleigh Club, and Sir Arthur Broke
was for many years its president. The first regular
meeting was at the "Thatched House," on February
7th, 1827, when Rennell's great friend, Mr. Marsden,
was in the chair. The navy was very strongly represented in the club by Captain Beaufort the
hydrographer, Basil Hall and Marryat, Franklin and
Parry, Smyth, Mangles, Cochrane, Murray, Mansell,
Beechey, and Owen; and the army by Colonel
Leake, Major Keppel (afterwards Earl of Albe-
marle), and Roderick Murchison. Other well-
known travellers were Baillie Fraser, Colebrooke,
Cam Hobhouse, Mountstuart Elphinstone, Sabine,
John Barrow, W. R. Hamilton, and Sir George
Staunton.

At the first dinner Sir Arthur Broke presented a
haunch of reindeer venison from Spitzbergen, a jar of
Swedish brandy, rye-cake baked near the North Cape,
a Norway cheese, and preserved cloud-berries from
Lapland. It was then agreed that each member

M 2

should be invited "to present any scarce foreign game, fish, fruits, wines, etc., as a means of adding greatly to the interest of the dinners, not merely from the objects of luxury thus afforded, but also for the observations they will be the means of giving rise to." The evening passed with the greatest enjoyment; and at the next dinner Captain Mangles presented some bread made from wheat brought by him from Heshbon, on the Dead Sea. Sir Arthur Broke contributed a brace of capercailzie from Sweden. At another dinner a ham from Mexico was presented by Mr. Morier, whose health was accordingly drunk. Thus the most eminent travellers in London were brought together, an interchange of ideas frequently took place, and the feeling that the creation of a more completely organized institution for the advancement of geography was necessary gradually took a definite shape. The Raleigh Club had freshened up old memories, had kept alive an interest in geographical pursuits, and had prepared the way for more systematic work. The Club continued to flourish after the formation of the Geographical Society, becoming more and more closely connected with it until, in 1854, the affiliation became complete, and the name of Raleigh was changed for that of the Geographical Club.

The question is a little complicated, but the evidence in favour of Admiral W. H. Smyth having been the founder of the Geographical Society is, on the whole, conclusive. Mr. Galton, in 1877, said that to Admiral Smyth " was due just one-half the credit of the foundation of the Society, which was established by the combination of two contemporary and independent schemes, of one of which the admiral was the sole originator." * But Admiral Smyth had sketched out his well-conceived scheme for a Geographical Society, and had enrolled many names in the very beginning of 1830, while the other scheme was not brought forward until the spring of that year. He, therefore, clearly has the honour of priority as a founder.

Major Rennell's death took place in March, 1830, between the births of these two schemes for the foundation of a Geographical Society. Admiral Smyth was Rennell's intimate friend, and they habitually exchanged views and opinions on geographical subjects. It is therefore almost certain that Rennell had been consulted by his friend, and that the last subject, or one of the last subjects, that interested the dying *savant* was the project of founding the Geographical Society.

The second scheme was set on foot by Sir John Barrow, the Secretary of the Admiralty, who summoned a meeting of the Raleigh Club on the 24th of May, 1830, which was numerously attended. It was unanimously decided that a Geographical Society was needed, and that its objects should be to print geographical information for its members, to accumulate a library and a collection of maps and charts, to procure instruments for the information and instruction of travellers, to prepare instructions for explorers and give them pecuniary assistance, to correspond with similar Societies and with geographers in all parts of the world, and to open communication with all philosophical and literary Societies with which geography is connected. This meeting took place just two months after Rennell's death.

---

* *R. G. S. Journal*, Vol. XLVII., p. 128.

A committee was appointed to arrange details and report to a second meeting, consisting of six members of the Raleigh Club. Sir John Barrow was the chairman, and the meetings took place at his room in the Admiralty. Dr. Robert Brown, " Princeps Botanicorum," was naturalist to the Australian expedition of Flinders, and had contributed the botanical appendices to Parry's " Voyages," Salt's " Abyssinia,' and Clapperton's " Journeys." He was secretary, president, and for many years the mainstay of the Linnean Society. Roderick Murchison, as a subaltern in the 36th, had served in the battles of Rorica, Vimiera, and Coruña, but retired in 1815. In 1830 he was a rising geologist. John Cam Hobhouse, the friend and companion of Byron, and an ardent liberal politician, always took an intelligent interest in geography. Mountstuart Elphinstone had been Governor of Bombay, and was known as a good historian and a sound geographer. Bartholomew Frere, the brother of Canning's intimate friend Hookham Frere, was a diplomatist and a well-read geographer and scholar. These six men formed a committee, to which the name of Admiral Smyth was added as soon as it was known that he had already commenced work in the same field. The two schemes were thus amalgamated, and a second meeting took place on July 16th, 1830, when twelve resolutions were passed, including the election of the first president and council. Sir John Barrow announced that four hundred and sixty names had been enrolled in the list of fellows : a proof that a favourable opinion had been formed of the utility likely to result from the labours of the Society. In the first council are to be found the names of Sir Arthur Broke, founder of the Raleigh Club ; of

Admiral Smyth, founder of the Geographical Society ; of all the members of the committee, except Frere and Murchison ; of Captain Beaufort the hydrographer, and Captain Horsburgh the East India Company's hydrographer; of Sir George Murray and Lord Prudhoe; while the vice-presidents were Sir John Franklin, Sir John Barrow, Mr. Greenough, and Colonel Leake.   Mr. W. R. Hamilton and Captains Basil Hall and Mangles, F. Baily the astronomer, and J. Britton the antiquary, complete the list of the more important members.   Of these, Sir John Barrow, Sir George Murray, Mr. W. R. Hamilton, Mr. Greenough, and Admiral Smyth were future presidents.   All the leading politicians and scientific men of the time, forty-four naval officers (including the king), and fifty officers in the army joined the Society.   Thus this new organisation, so urgently needed, and destined to work so much for good in the prosperous future that was in store for it, auspiciously commenced its career.

King William IV. had taken a great interest in the works of Major Rennell, as his father had done before him ; indeed, George III. gave considerable aid in their publication.   As Lord High Admiral, the sailor-prince received from Lord Spencer an urgent representation with respect to the importance of publishing the current charts ; and, as William IV., he accepted the dedication from Lady Rodd.   He readily consented to become the patron of the new institution, desiring that it should be called the Royal Geographical Society.   The king also granted an annual donation of fifty guineas to constitute a premium for the encouragement and promotion of geographical science and discovery, saying to Sir

Herbert Taylor that he hoped his own officers would often win it. Until 1838 the whole premium was given to one recipient, but when her present Majesty intimated her intention of continuing her uncle's grant, it was divided into two awards of equal value —the Founders' and Patron's Gold Medals—adjudicated annually to two recipients; and in that form the royal premium has continued to be bestowed ever since, except in 1850 and 1855, when only one medal was granted.

The African Association was merged in the Geographical Society in 1833, adding a sum of three hundred and eighty pounds to the funds; and Mr. Bartle Frere, as its representative, became a member of the Council. The Palestine Association, which had been formed early in the century, also resolved that its funds, amounting to one hundred and thirty-five pounds, papers, and books, should be made over to the Society, to be employed as the Council might think fit for the promotion of geographical discovery.

Numerous old and intimate friends of Major Rennell joined the infant Society, including Lord Spencer, Marsden, Leake, Barrow, Smyth, Beaufort, Franklin, Parry, Mangles, Inglis; while his son-in-law, Admiral Sir J. Tremayne Rodd, became a member of Council and a considerable benefactor to the library. His grandson, James Rennell Rodd, was a Fellow of the Society from 1830 until his death, in 1892.

For the first ten years of the Society's existence, it was, through the kindness of Mr. Robert Brown (Princeps Botanicorum), housed in the rooms of the Horticultural Society, in Regent Street. From 1840 to 1854 it rented rooms at 3, Waterloo Place, and the meetings were held at King's College, by permission of the Principal. The house at 15, Whitehall Place, was rented from 1854 to 1870, meetings at first being held in the large library ; but for twelve years—from 1858 to 1870—they were held in the fine room at Burlington House, on the left of the court, now pulled down. At the termination of the lease at Whitehall Place, the freehold of 1, Savile Row, was purchased, where the Society has ever since been established, and for the last quarter of a century the meetings have been held in the neighbouring hall of the London University, by permission of the Senate. From small beginnings, the Society has increased, during the sixty-four years of its existence, to 3,775 Fellows, with an invested capital worth twenty-seven thousand pounds, and an annual income of ten thousand pounds. The Society has spent twenty thousand pounds on exploring expeditions, and large sums in furthering geographical education. It has a system of instruction for geographers and explorers, and disseminates information throughout the world. Since 1855 the Government has granted an annual sum of five hundred pounds in order that the collection of maps might be rendered available for general reference ; and the Society has now become a great national institution.

Such has been the machinery through which the work of Rennell has been continued and perfected. Commencing in the very year of the great geographer's death, it is alike a monument to his memory and a witness to the value of his labours; for in almost every department it will be seen that the foundations were laid by James Rennell.

The atlas of Bengal and the map of Hindostan, with its memoir, are the foundations on which the splendid surveys of our Indian Empire have since been built. For many years these maps of Rennell were the chief, almost the only, geographical authorities for India. It was not until 1802 that Major Lambton's proposal for a measurement of an arc of the meridian and the execution of a trigonometrical survey of India took shape, the longitude of the Madras Observatory being adopted as a secondary meridian, and a first base-line being measured near Madras, and a second near Bangalur. Lambton's labours in the field were of a most arduous character ; and he was called upon from time to time to demonstrate the utility of his work. Even Major Rennell at first expressed the opinion that route surveys on an astronomical basis were equally accurate and more economical ; but he became convinced of the superiority of Lambton's method long before his death, and welcomed with joy the six-sheet map of India by Walker, which was partly based on triangulation. The importance of the survey was not fully recognised by the Government until 1818, when the Governor-General ordered it to be denominated for the future " The Great Trigono-metrical Survey of India." Colonel Lambton died in 1823, worn out by incessant toil under a tropical sun, and was succeeded by his zealous and accomplished assistant, George Everest. Everest was in England from 1825 to 1830, studying the newest improvements and superintending the construction of instruments on the most improved principles, including compensation bars for the measurement of bases, instead of the old inaccurate method by chains. Everest returned to India in 1830, combining the appointments of Superintendent of the Trigonometrical Survey and Surveyor-General in his own person. Thus far the progress of the work initiated by Rennell was personally watched by that father of Indian geography. The year in which Rennell died saw the foundation of the Geographical Society, and the return of Everest to Calcutta.

From that date, during the next sixty years, the magnificent edifice of the Indian Surveys has been steadily built up by a succession of the most zealous, accomplished, and resolute geodesists that Britain or any other civilised country has ever produced. In 1831 the first base-line was measured with the compensation bars at Calcutta. Everest entirely altered the old system of his predecessor by substituting the gridiron for the net-work method. He introduced the compensation bars, invented the plan of observing by heliotrope flashes and the system of ray tracing, and designed the plan for the towers. He also planned a complete revision of the famous Bengal Survey of Rennell, which had held its own for fifty years. Great progress was made during Sir George Everest's term of office—from 1823 to 1843; and although there have been modifications and improvements since his time, nearly everything in the surveys was originated by this great geodesist.

On the retirement of Sir George Everest, his able and indefatigable lieutenant Andrew Waugh succeeded him, and set steadily to work to complete the scheme of triangulation, while topographical and revenue surveys were pushed forward at the same time. The North-Western Himalaya series, executed between 1845 and 1850, was the longest between measured bases in the world, being one thousand six hundred and ninety miles long. From this series the heights of seventy-nine Himalayan peaks were

measured, including Mount Everest, the highest of
all: 29,002 feet above the sea. I have dwelt
elsewhere on the dangers and difficulties incurred
in the execution of these surveys, which are quite
equal to those encountered in the majority of Indian
campaigns.* The Indian surveyor devotes great
talent and ability to scientific work in the midst of
as deadly peril as is met with on the battle-field, and
with little or no prospect of reaping the reward that
he deserves. His labours are of permanent and lasting
value, but few know who obtained these valuable
results except his immediate chief and his colleagues.
Sir Andrew Waugh's principal undertaking, besides
the completion and extension of the triangulation
designed by Sir George Everest, was the survey of
Kashmir and the mighty mass of mountains up to
the Tibetan frontier, under the superintendence of
Colonels Montgomerie and Godwin Austen. The
latter sketched some most difficult ground with
great taste and skill, including several enormous
glaciers; and a peak was measured, provisionally
called K. 2, which is 28,290 feet above the sea.
It is now known to geographers as Mount Godwin
Austen.

The successor of Sir Andrew Waugh, who retired
in 1861, was General James T. Walker, while Sir
Henry Thuillier became Surveyor-General. Topographical surveys were continued in all parts of
India; the trigonometrical work was practically completed, and many surveyors penetrated into Persia
and Central Asia, extending the knowledge of the
countries bordering on British India. The brilliant

---

* "Memoir on the Indian Surveys" (second edition), p. 104.

work of the surveyors in the service of our Indian
Government and the successors of Rennell has received cordial recognition from the Geographical
Society. Gold medals have been adjudicated to Sir
Andrew Waugh, Colonel Montgomerie, and Colonel
Holdich, while twenty explorers in Central Asia,
including four natives, have been granted medals or
minor awards; so that from the foundations laid by
Rennell a very noble scientific edifice has been raised
by the surveyors—his successors—in the sixty years
following his death.

In the study of historical and comparative geography the succession from Rennell has not been
so complete or so continuous. Yet Western Asia,
between India and the Mediterranean, has been a
field of investigation on which the gold medal of
the Geographical Society has been won six times
since the death of Rennell: by Rawlinson, and
Chesney, Layard, Hamilton, and Symonds, who
were all occupied on the same investigations that
occupied the thoughts of Major Rennell during a
long course of years. But Sir Henry Yule and Sir
Edward Bunbury are the successors who come
nearest to Rennell in their methods of research, and
in the interest they have taken in reading ancient
history by the light of modern geography. Yule,
like Rennell, had the advantage of getting his
geographical training in the field, and in the very
same corps to which Rennell belonged, the Bengal
Engineers. Returning home in 1862, Yule brought
many years of experience in the East, a great fund
of knowledge amassed from books, and a fine critical
instinct to bear on the literary tasks he undertook.
The chief of these were "Cathay, and the Way

Thither," issued by the Hakluyt Society in 1866,
and his edition of " Marco Polo." The Hakluyt Society
had been established in 1847 for printing rare and
unedited voyages and travels. The work of such a
Society was after Yule's own heart. He contributed
six volumes to its series, and he was president from
1877 until his death, in 1890. Sir Henry Yule's
wonderful memory, his passion for accuracy and
thoroughness (insisting upon an index being properly
prepared for every volume), and his love for old
travellers and their narratives, made him an ideal
president for such a Society. He wrote a memoir
of James Rennell for the *Royal Engineers' Magazine*
with that painstaking accuracy, with occasional
touches of humour, which characterize all his
writings; but it was all too brief, although he contemplated writing a fuller biography: a design which
he was not spared to fulfil. Like Rennell, Sir Henry
Yule was made an Associate of the French Academy;
but he only received the honour just before his death,
while Rennell enjoyed it for many years. The last,
or almost the last, letter Sir Henry Yule ever dictated
was one of thanks for this much-coveted honour;
and he concluded it with these touching words,
" Cum corde pleno et gratissimo moriturus vos saluto."
He received the gold medal of the Royal Geographical
Society in 1872 for the eminent services he had
rendered to geography in the publication of his three
great works.

Sir Edward Bunbury was perhaps more closely
in sympathy with Major Rennell than Sir Henry
Yule, from the point of view of his Western Asiatic
studies; for Sir Edward had occasion to read the
work of Rennell on the "Geography of Herodotus" with

care, and to criticise some of his conclusions, in the preparation of his own great work on "Ancient Geography." Like Rennell, Sir Edward is remarkable for thoroughness and accuracy; and the late Professor Freeman used to say that when he saw E.H.B. at the end of an article he always felt perfect confidence in the correctness of the information it contained. His "Ancient Geography," compendious and exhaustive, and at the same time written in a scholarly and agreeable style, will, as Sir Henry Rawlinson anticipated, be found in the library of every geographer and scholar, and become a text-book in the educational establishments of the country.

But Sir Henry Yule and Sir Edward Bunbury, to whom must be added the honoured name of Sir Henry Rawlinson, are the only great successors of Rennell in England, as comparative geographers. This interesting and very important branch of the science has been comparatively neglected in this country. The foundation of the Hakluyt Society indicated that the subject was not without interest for English readers; and the work of Major, Burnell, and some others, was excellent of its kind. Rylands and Schlichter have lately shown us that the study of Ptolemy still has attractions for geographers of this generation; and the Geographical Society's last travelling student, Mr. Beazley, has turned his attention to the identification of the places mentioned in the Periplus, of the Erythræan Sea, and to a revision of the work of Dr. Vincent. It is to the institution of the travelling studentships of the Royal Geographical Society, in conjunction with the Universities of Oxford and Cambridge, and to the geographical readerships at our principal seats of learning, that we

must look as the most hopeful agencies for the
revival of a taste for comparative geography. Mr.
Mackinder and others have recently explained to
students with great force and ability the close connection between geography and history. Until this
connection is impressed upon the rising generation
of travellers, and is present in their minds, we shall
continue to be deluged with rubbish in the form of
books of travels, and we shall look in vain for the
charm which is felt in reading the works of travellers
who were also scholars and observers. To such men
a distant country is not a mere succession of mountains and plains, where more or less wild sport is to
be found. To them every hill and dale is alive with
associations of the past. They derive lasting pleasure
and instruction from their impressions and investigations, and they impart them to others in the most
useful and interesting form that a geographical
narrative can take. They also furnish really valuable
material for students. We must look to the labours
of our Mackinders and Yule Oldhams for the develop-
ment of a new generation of Rennells and Vincents,
of Yules and Rawlinsons, and for the revival of that
taste for geographical research which has hitherto
received less real encouragement among us than in
almost any other country of Europe.

Major Rennell's work in connection with the
African Association was not the least valuable of his
services. In the second volume of his "Geography of
Herodotus" he laid the foundations of the study of
the geography of Africa; and by compilations of maps
containing all existing information, and by his elucidations of the travels of Mungo Park and others,
he became the real centre of African exploration

during a long term of years. As regards actual discovery, the efforts of the Association were mainly concentrated on the Niger, although it originated good work both on the Nile and in the Libyan desert. Rennell lived just long enough to see the solution of the Niger question, and in a way which he had not expected.

Thus the Geographical Society, on its creation, found the Niger problem just solved. The practical result of the Niger exploration has been that a new market is opened for British manufactures, now under the auspices of the chartered Niger Company : that a very considerable addition has been made to the sum of human knowledge : and that the world is the better for many records of bravery and heroic devotion to duty.

Many desultory African expeditions were encouraged and some were assisted by the Geographical Society during the first twenty years of its existence; but it was not until Dr. Livingstone appeared on the scene that the era of discoveries on a large scale commenced. In this department of geographical work, the mantle of James Rennell fell on the shoulders of Sir Roderick Murchison, who, among his other great services, was the steady promoter of African discovery, and the warm friend of African travellers during twenty years. As Rennell's map of Northern Africa inaugurated the operations of the Association, so Sir Roderick's address in 1852 on the " Great Features of Africa " may be considered to have commenced the modern era of discovery. Livingstone's famous journey down the valley of the Zambezi was made between 1853 and 1856, and from 1858 to 1864 he discovered the Shiré and Lake Nyassa. In the same period Dr. Barth was

N

making excursions round Lake Chad, discovering the great river Benuè, and completing a hazardous and adventurous journey to Timbuktu. In another direction the equatorial lakes of East Africa and the sources of the Nile were being made known. That accomplished scholar and traveller, Richard Burton, discovered the Lake Tanganyika in 1858 ; and in 1862 Speke settled the question of the Nile sources. The problem of ages had occupied the attention of Major Rennell during his studies of Herodotus, Ptolemy, and Edrisi, and it had excited his interest in a high degree. He had not ventured to place the sources of the Nile further south than 6° N.; but Speke's discovery established the fact that the most distant Nile sources are well south of the equator. In the thirty years that have elapsed since the death of Speke, the great work of completing the discovery of the interior of Africa has steadily progressed. Livingstone, Burton, and Speke all received large grants in aid, as well as encouragement and cordial sympathy, from the Council of the Geographical Society ; and in 1877 that body resolved to raise a special " African Exploration Fund ": a measure which resulted in the despatch of the expedition led by Joseph Thomson, and organized from first to last under instructions from the Council. It had a clearly defined aim, and was conducted ably, economically, and with complete success. The Council has shown its appreciation of their work by adjudicating the royal medals to twenty-four African travellers : the first to Lander, in 1832, for discovering the mouth of the Niger ; the last to Selous, in 1893, for twenty years of pioneering work in Mashonaland. There have also been seventeen minor awards conferred for work in Africa. The Geographical Society may fairly claim to have carried on the work bequeathed to them by Major Rennell and the African Association with great liberality, with ability, perseverance, and success.

Major Rennell was the founder of oceanography, or that branch of geographical science which deals specially with the ocean, its winds and currents; to which, in recent years, has been added the study of its depths. His current charts and memoir, constructed and prepared with extraordinary diligence and ability, are the foundations upon which all subsequent investigations have been based. He was assisted and advised by accomplished marine surveyors, such as Smyth and Beaufort. The former was the founder and president of the Geographical Society. The latter was hydrographer for a quarter of a century, and on the Council of the Society for twenty years, always maintaining the most cordial relations between the Society and his department at the Admiralty. His successor, Captain Washington, was secretary to the Society and also on its Council; and the three subsequent hydrographers have been vice-presidents, and almost continuously on the Council. An annual report from the hydrographer on the progress of the surveys always follows the president's address; and indeed, their aid has been so continuous since the foundation of the Society, that the hydrographers have almost become *ex-officio* members of the Council.

Mr. Findlay, who served on the Council of the Geographical Society for nearly twenty years, from 1857 to 1874, was perhaps the most direct descendant of Major Rennell's hydrographical side, for he was the successor of John Purdy, the personal friend of Rennell, and editor of his current charts. Purdy died

N 2

in 1843; and in 1851 Mr. Findlay published his well-
known directory for the navigation of the Pacific
Ocean. He devoted several years of intense labour
and application to this work, which was the model of
all his later productions. These comprised a series of
six nautical directories for the whole world, which are
monuments of industry and perseverance, and standard authorities in every quarter of the globe. Findlay,
like Rennell, took a deep interest in Arctic exploration.
He was one of the leading advocates for its renewal in
1875, and just survived to see the expedition of Sir
George Nares fitted out. As the successor both of
Purdy and of Laurie, who gave Rennell so much
useful aid in his hydrographical researches, Findlay
was better fitted than any other man to write the
paper on ocean currents which was published in the
Geographical Society's Journal for 1853. He there
declared that, although detached facts and numerous
observations had been recorded, yet the generalisation
of these data, and their reduction to a uniform system,
remained nearly in the same state as when Major
Rennell completed his investigation of the currents
of the Atlantic. In other words, the hydrographic
work of Rennell held the field without a rival for a
quarter of a century after his death. The very first
contribution to the science of oceanography was
Rennell's memoir on the Agulhas current in 1778, so
that he was the father of oceanography. His subsequent memoirs contained the first elucidations of
the system of oceanic circulation; and his current
charts, published under the editorship of Mr. John
Purdy in 1832, embodied all previous knowledge.
The invention of apparatus for deep sea sounding has
opened out a whole world of investigation in the

depths of the ocean of surpassing interest. The science has grown in the last forty years with rapid strides, and now we are on the eve of the completion of the monumental work of the *Challenger* expedition. But Rennell ought always to be remembered, and his name should be held in honour, as the originator and founder of this branch of geographical science.

In this chapter a glance has been cast over the progress of the several departments of the science of geography in which Major Rennell took the deepest interest, since his death, sixty-four years ago. This is, in fact, a review of the work of the Geographical Society, which was founded in the year that he ceased to live and to work, and has carried forward his labours with zeal and perseverance. The Society is his successor and executor. How great the debt of posterity is to the illustrious geographer cannot be measured. The information he collected, his generalisations, his methods, and his treatment of doubtful problems, became the common property of his fellow-countrymen. They have permeated more or less the writings and thoughts of his successors, and must have exerted a subtle, but important influence on succeeding generations of geographers. All cultivators of that popular and most interesting science of geography owe a debt which can only be paid by remembering the high qualities and devoted zeal, and by striving to imitate and follow the noble example of James Rennell, who, so far as this country is concerned, was the founder of their science.

CHAPTER XI.

MAJOR RENNELL was a great sufferer from the gout
during the last years of his life. He lost the use of
his hands during long intervals, and was often confined to his bed or to a sofa. But his daughter,
living in Wimpole Street, was a constant visitor,
and her children, of whom he was very fond, were
frequently with him. He continued to hold his
forenoon receptions, and maintained a keen interest
in public affairs and in those geographical pursuits
which had formed the chief work of his life. Some
old friends had been removed by death. Sir Joseph
Banks had passed away in 1820. But Rennell had
formed pleasant intimacies with younger men.
Always a sailor at heart, his later friends were for
the most part naval, including Beaufort and Smyth,
Franklin and Parry, and his son-in-law, Tremayne
Rodd.

The great English geographer had been an
Associate of the Institute of France since the 26th
of December, 1801. In 1825 the gold medal of the
Royal Society of Literature was awarded to " Major
Rennell, who for half a century had been pre-eminent
among the geographers of Europe." Owing to his
infirmities, he could not attend at the Society's rooms
in Lincoln's Inn Fields to receive it. A deputation,

consisting of the President, the Bishop of St. David's, Archdeacon Nares, Sir William Ouseley, and others, therefore waited upon Major Rennell at his own house in Nassau Street (formerly Suffolk Street) where he received from the bishop's hands that honourable testimonial of literary merit.

Rennell was pleased by this recognition of his work. Writing to Captain Beaufort, he said: " I am truly thankful for your kind congratulations. These are the rewards of such people as ourselves, who do not place the *summum bonum* in dross and dirt, but in the good opinions of our fellow-citizens "; and to Mrs. Smyth :—" Your good wishes and congratulations are worth a dozen medals. I confess I was much gratified by the adjudication of the medal, and felt myself much honoured by the mode of presenting it to me." His labours were fully appreciated by his contemporaries; and although they are less known to later generations, their influence is still felt among us in many ways.

Rennell's correspondence in his old age is occupied with interesting subjects pleasantly treated, and every now and then a phrase or sentence reminds us of the writer's education at sea and of his habit of looking at things from a sailor's point of view. Thus, in referring to Captain Basil Hall's book, he writes :—" There is much taste and judgment in the editorship. He sets us down at a place (sleeping by the way, like Ulysses), and having the hawsers in the tier for those who are accustomed to warping." His letters to Captain Beaufort relate to the literary subjects which interested both, such as the paper on Jerash, and questions relating to Levanters in the Mediterranean and the wind *Euroclydon.* They exchanged and lent

new books to each other—for it was long before the
days of Mudie—and discussed them in their letters.
The " Naval Histories " of James and Brenton appeared
at about the same time, and Rennell's opinion was
that "neither of the ' Naval Histories' was what it
ought to have been; but I prefer James', as being the
more useful of the two. The critique on Brenton in
the *Edinburgh Review* is much too personal to pass
for criticism, but I pity not the man who seeks to
degrade Lord Howe—as great a man as Nelson, but
not always fortunate. Nelson owed much to Lord
Howe's moral as well as physical victory. I never
knew what Nelson's idea of Lord Howe was, but
heroes always respect each other. I was much pleased
with James's account of Trafalgar, and have studied
it; but more is wanting about Lord Howe's 1st of
June."

In these later years Major Rennell took a deep interest in the Arctic expeditions which were despatched
between 1818 and 1827. He was intimate with Sir
John Barrow, as well as with Parry and Franklin,
and his friend Beaufort was appointed hydrographer
in 1825. When Franklin's narrative of his terrible
sufferings in the frozen lands appeared in 1823,
Rennell declared that he had never heard of such
trials before or of such heroic fortitude in facing
them, and he looked forward to discussions on the
subject with Beaufort and Franklin. He had read a
long letter about the Mackenzie River from Kendall
to Mr. Walker, of the Admiralty, and suggested that
the great body of fresh water flowing down it might
be one cause of the absence of ice near its mouth.
In 1826 Franklin was away again on his second
expedition, and we find Rennell thanking the hydro-

grapher for his kindness in allowing him the perusal of the explorer's interesting letters. "Franklin seems full of hope, and in the mood that promises success based on reason and good sense. Poor man! how I felt for him when I read that part of his letter where bitter reflection returns; and I was very much touched when I found that he remembered me with so much kindness in the midst of the desert. The letter is quite a treat throughout. I never knew a man so full of hope—founded, I think, on rational expectation. I am much gratified to be remembered on the borders of the Arctic Circle."

Major Rennell took the same interest in the attempts of Parry to make the North-West Passage, once by Hudson Strait and twice by Lancaster Sound; but he was not sanguine of success in the direction of Prince Regent's Inlet. Writing to Mrs. Smyth on October 31st, 1825, he said:—"I have to thank you for your kindness in communicating to me tidings of the safe return of our adventurous navigator, Captain Parry. Success in his attempt I could not expect; all that I looked forward to was his safety. It was my opinion, as he knew, that the Regent's Inlet would lead him into a *cul-de-sac* of packed ice at the back of Melville Peninsula: in fact, the ice which Captain Franklin saw continually drifting to the eastward with the prevalent westerly winds through 'Franklin's Pond,' as I call it. Accumulating in the bay behind the Fury and Hecla Strait, it would attach itself to the ice with which Captain Parry saw it filled up during his previous voyage." This was the last voyage to the north during Rennell's life. But his name had already been immortalised in the Arctic regions by his friend Captain Parry during the first voyage.

Immediately to the eastward of Cunningham Inlet, on the coast of North Somerset, there is a bold headland. It was clearly distinguished on August 29th, 1820, when the *Hecla* was on her return voyage; and Parry wrote in his journal:—" I named the headland after Major Rennell, a gentleman well known as the ablest geographer of the age." *

The last subjects of importance which occupied Rennell's attention were these Arctic geographical problems, and the discoveries of Captain Clapperton in the basin of the Niger. Parry's observations of currents and sea temperatures during the second voyage were valuable as material in the construction of Rennell's Atlantic current charts; while the light thrown on the geography of the far north by Franklin and Parry would further have assisted the illustrious geographer in his generalisations if it had reached him earlier. But the close of a long and valuable life was approaching.

Severe physical suffering was assuaged by the pleasure of conversing with friends, by watching the gradual solution of questions which had long occupied his mind, and, above all, by domestic affection. The attentions of Lady Rodd never flagged, and her children were an abiding interest to the sufferer especially his one little grandson and namesake— "double namesake," as he called the child—James Rennell Rodd. He was born on February 28th, 1812, and his three sisters followed, and were named Alicia, Frances (born in 1815), and Wilhelmina. The letters of Major Rennell to his grandson cover the period from 1812 to 1819, and they are so charming, so

* " Parry's First Voyage," p. 265.

sympathetic, and enter so completely into all that would
interest a young boy, that some extracts from them
cannot fail to be acceptable to those who have followed
the life-story of the great geographer. Rennell always
seeks to raise the thoughts of the boy he loved with
such tender affection, while he interests and amuses
him. His letters are full of lively sallies and bright
and witty passages, but he never resorts to injudicious
banter. He was past eighty and the boy was twelve
—there were seventy years between them—yet the
venerable and learned *savant* always treated his
grandson with deference, anxious that no words of
his should hurt the boy's feelings or wound his sus-
ceptibilities. When the child was five and six the old
major printed the letters in red ink. In due time
young Rodd was sent to Dr. Pinckney's famous school
at East Sheen, and his grandfather went down to pay
him a visit and walk round the playground with him.
A few days afterwards he wrote and told him how
much he had thought of him and of his school, and
that he had derived much comfort from what he saw
there. " You have only as yet skimmed the surface
of English history," he went on. " You have not
perhaps read of Sir William Temple, a great states-
man during the time of King William III., and who
indeed, to the honour of the king's taste, was admitted
to his private society and friendship. Your present
school was then the country house of Sir William
Temple, and there (as well as at Moor Park, another
of his houses) King William was accustomed to visit
Sir William as a friend, and their conversation was
supposed to relate in great measure to State affairs—
that is, government. Kings have not at all times
sought such kinds of companions, but King William

was a wise and an honest man, and, I verily believe,
acted for the best. Now you must think, when you
are playing under that magnificent tree which stands
in front of the house, that perhaps King William and
Sir William Temple have sat in conference under the
very same tree, revolving in their minds the best
means of securing the comfort and happiness of
millions of people. And you must also bear in mind
that it was Sir William Temple's high personal
character, added to his talents and knowledge, that
was the cause of this connection, so honourable
to both. Sir William's mansion has had a fate
of a more dignified and useful kind than many
others that were formerly inhabited by some of our
great characters. Mr. Pope mentions one or two of
them which had passed to money-dealers and city
knights. Now yours has become a nursery from
whence the future Temples are to be transplanted,
and finally destined to govern, ornament, and enrich
their country.

"I received, in the presence of your dear father,
your very affectionate letter congratulating me on my
arrival at the entrance of my eighty-first year. I
thank you most kindly for your affectionate attentions to me, and trust, by the blessing of God, that
I may escape what David has so justly said is so
commonly the .lot of those who are said to 'be so
strong'; nor have I a wish, whilst I preserve my
faculties, that it should so soon pass away. I am
truly sensible of the great blessings which I enjoy,
and, not the least, that of having so good a grandson.
I rejoice to think that we are so soon to meet. Pray
present my best respects to Dr. Pinckney."

It is curious how one train of thought led to

another quite different in these letters to his young descendant, showing how completely the old man abandoned himself to the pleasure of free communion with one whose age brought back all the delightful memories of his own boyhood. " You say," he wrote, " that you are reading 'Homer' with your mamma You must be greatly delighted. Mr. Pope's poetry has so much harmony in it. He seems to have spun verses as freely as a spider his web;" and this figure leads Rennell to thoughts of the huge spiders he had seen during his wanderings on the Island of Rodriguez when he was a midshipman, and he gives his grandson a most interesting account of them. He then mentions the entertainment to which he went, given by Captain Parry, on board the *Hecla*.

When his grandson had reached the age of fourteen, and had gone to Eton, we find Major Rennell answering his questions about the shape of the globe, and explaining why the feeling of cold is greater when the wind is blowing than in a calm, and he tells him much home news. " I dined with papa and mamma on Tuesday (February, 1825), and met with very agreeable company. There were two persons — Lord Carleton and Lord Stowel—one of eighty-five or eighty-seven, and the other about my age." In another letter of the same year there is an account of Clapperton's African journey, and of Belloo, King of Sokatu, written carefully, in a way that would be perfectly clear to a boy of that age. In the following year young James began Xenophon's " Anabasis," and his grandfather reminded him that this was the book upon which he had written, in order to illustrate its geography. " You will like it when you are able to comprehend it fully. I take Xenophon (take him as they say, 'all in all') to be one of the greatest men of antiquity."

From 1825 to 1829 James Rodd was at Eton, and continued to receive these charming letters from his grandfather : one a most interesting and instructive one on the diverse habits of the ostrich and cassowary. The last letter is dated the 1st of December, 1829, written with gout both in the hands and feet, but, like all the others, cheerful, affectionate, and instructive. Major Rennell was now eighty-seven years of age, and in the last month of 1829 he slipped from an arm-chair and broke his thigh. He hardly ever left his bed again, and died on the 29th of March, 1830.

There was only one suitable resting-place for the remains of the greatest geographer that England has produced. On the 6th of April they were interred in the nave of Westminster Abbey, and there is a tablet to his memory, with a bust, in the north-west angle of the nave. On the day after the funeral the *Times* published a notice, from which the following is an extract :—

"One quality which peculiarly marked his writings, and which cannot be too much held up for imitation, is the ingenuous candour with which he states the difficulties he could not vanquish or acknowledges the happy conjectures of others. Those who have studied his 'Geography of Herodotus,' and followed under his guidance the ' Retreat of the Ten Thousand,' will have felt how much this quality augments the value of his reasonings. In all his discussions his sole object was the establishment of truth, and not the triumph of victory. Another characteristic of this amiable philosopher was the generous fidelity with which he imparted his stores of learning in conversation. A memory remarkably tenacious, and so well arranged as to be equally ready for the reception or the distribution of knowledge, made him a depository of facts for which few ever applied in vain. Adapting himself to the level

of all who consulted him, he had the happy art of correcting
their errors without hurting their feelings, and of leading
them to truth without convicting them of ignorance."

Till Rennell's time, as Sir Henry Yule truly observed, it could hardly be said that England could
boast of any great geographer. His pre-eminence in
that character is still undisputed, like that of D'Anville
in France, and of Ritter in Germany; but the superiority of Rennell over the French and German students
consists in his many-sidedness. D'Anville and Ritter
were students, and nothing more. Rennell had far
wider experiences. He was a sailor and a marine
surveyor, accustomed from boyhood to active service,
and habituated to the study of nature in many seas.
He was a land surveyor and explorer, mapping the
vast ramifications of the deltas of the Ganges and
Brahmaputra, and exploring the wild forest region at
the base of the Himalayas. It was his twenty years'
experience in the field, at sea and on shore, that
placed him on a far higher level than D'Anville or
Ritter. The beauty of his character, one trait in
which is referred to by the writer in the *Times*, is
also shown by his friendships and by his family
relations. Gentle, courteous, and simple-minded, he
was the type of a perfect English gentleman.

The best known portrait of Major Rennell is the
miniature by Scott, painted for Lord Spencer, which
has already been described. It was engraved by A.
Cardon, and published in 1799, appearing in the
European Magazine of 1802, and is reproduced in
the Frontispiece. An excellent porcelain medallion
was executed at Sèvres from a model, when the
major was much older. A replica was presented to

the Royal Geographical Society by Lady Rodd. There was another at the India Office, which Sir Henry Yule caused to be photographed to illustrate his article on Rennell in the *Royal Engineers' Journal* of January, 1882. A beautiful wax model by Hogbolt was presented to Sir Henry Yule by Major Rennell's only surviving grand-daughter in 1882. He left it to Sir Joseph Hooker in 1890, who presented it to the Royal Society. There is also a small portrait in uniform, painted in India when he was about thirty-five years of age, which is now in the possession of Mrs. Rennell Rodd.

The *éloge* on Rennell at the French Academy was read by Baron Walckenaer, the secretary, on August 12th, 1842. He said that in early youth Rennell's habit of reading the authors who wrote on geographical subjects aroused the desire to imitate them. He read with avidity the works of travellers and historians, and, not having had time to acquire a knowledge of Greek and Latin, he supplied this deficiency by studying the translations. He loved to familiarise himself with the rich literature of his own country, and it is due to this taste that he acquired the power of writing with purity in his own language. Rennell was so deeply impressed with the importance of hydrography for ensuring the prosperity of his country that he commenced his career with work in this branch of science, and he closed it while engaged on the same labours. Baron Walckenaer then refers to his seven years' service on the survey of Bengal, and remarks upon his retirement with the rank of major :—

" But that simple title, which one is accustomed to associate with the name of the English geographer, seems, from its use by him, to acquire a lustre superior to other titles. It is true

that the worship rendered to science produces effects similar
to those of a more venerable cult : it raises the humble charged
with good works, and debases the proud loaded with the vain
honours of the world."

After quoting the passage from the addresses of
Sir Joseph Banks, already referred to, with reference
to the " Bengal Atlas," Baron Walckenaer says that—

"Major Rennell had been received in England with an
empressement equal to the reputation he had made and to
the services he had rendered.   But it was owing to his social
qualities as much as to his talents that he made powerful
friends.   He was successively elected a Fellow of the Royal
Society, a Member of the Institute of France, of the Imperial
Academy of St. Petersburg, and of the Royal Society of
Göttingen.   If Rennell did not wish to accept either dignities
or wealth, it was not that he might obtain rest, but that he
might preserve his independence and devote himself entirely
to the projects he had conceived.   He aspired to a higher
renown than could be obtained by the publication of maps.
He desired, by his writings, to take his place among critical
geographers ; or, rather, he simply obeyed that passion for
geography which, when it has once got possession of the
intellect, grows until it is satisfied, and furnishes new means
of acquiring a more complete knowledge of the globe we
inhabit, the phenomena produced there, the productions that
are renewed, the people that have appeared and move on its
surface.   The thought, when it succeeds in maintaining itself
at such an elevation in time and space, no longer perceives the
succession of events and the conflicting interests, except at the
distance where history will one day place them."

After discussing the " Bengal Atlas " and the
" Memoir on Hindostan," Baron Walckenaer goes on
to say that the reception they met with is not explained altogether by their intrinsic merit or by the
interest caused by the political and military events of

o

which India was the theatre at that time. He held
that there was another cause, which it would be
useful to explain:—

"The history of the progress of geography in modern times,
down to the days of Rennell, is comprised in four or five
names. Ortelius and Mercator—rivals and friends—and afterwards the Sansons, had co-ordinated in a vast collection,
contained in voluminous atlases, everywhere copied, everywhere reproduced, the materials of the science. But Guillaume
Delisle was the first to cast off the heavy loads of error borrowed
from Ptolemy and the reveries of the Middle Ages. With the
aid of recent astronomical observations of itineraries which
had been transmitted to us by the ancients, Delisle founded
the system of modern geography. The edifice, of which he
laid the foundations, was completed and perfected by D'Anville.
Thus Belgium and France had alone produced men who were
distinguished as geographers. Those who were eminent as
followers in their footsteps—Blaeuw, the intelligent disciple of
Tycho-Brahé, Homann, Hasius, Wischer, Robert de Vagondy,
Phillipe Buache—were all Belgians, Germans, or Frenchmen.
The country of Isaac Newton, down to nearly the end of the
eighteenth century, had not given birth to a single geographer
who had made a name equal to the least of those above
mentioned. Major Rennell, by the publication of his maps
and his 'Memoir on Hindostan,' made that vacancy disappear
which existed in the ranks of the eminent men of all kinds
whom England had produced, and the satisfied sentiment of
national pride increased once more the number of eulogies
which were due to geography."

Baron Walckenaer then refers to Rennell's labours
in connection with the African Association, and to his
great works on the "Geography of Herodotus" and
the "Retreat of the Ten Thousand." The work on
the "Currents of the Atlantic and Indian Oceans"
occupied the last years of his life, and Baron
Walckenaer says that:—

"The 'Current Charts' of Rennell, and the explanatory volume which accompanied them, form the most learned attempt that had yet been made touching that department of the science. But the fact that he worked unceasingly to rectify his charts, is a proof that he himself was not entirely satisfied with them.

"However, he lived to a great age, though of a delicate constitution, which had been injured by wounds received in his youth. Sobriety, moderate daily exercise, the recreations of society after the hours of work, tender care taken of him by his family, produced this happy result. It ought not to be omitted, also, that the excitements of ambition and of politics never troubled either his days or his nights. Not that he was indifferent to what concerned the welfare of his country, nor a stranger to the disagreements of those who shared the responsibilities of its government. A friend of Fox and of Lord Spencer, he belonged to that party which believed that the English Constitution ran more danger of being injured by the encroachments of the Crown than by the invasions of parliamentary authority. He was a Whig in the older sense of that word. But when he was consulted on subjects relating to his special studies, he showed an equal zeal in enlightening all those who were capable of profiting by his information for the advantage of his country, to whatever party they belonged."

Baron Walckenaer concludes his *éloge* by giving some account of the personal appearance of his hero and of his last illness. It is a noble tribute, and all the more valuable because it is the production of an absolutely impartial *savant* of great learning, and upon whose judgment the most implicit reliance may be placed. Moreover, it becomes still more gratifying when we remember it is the production of a Frenchman—the native of a rival country. But where geography is concerned, the rivalry between France and England has never been anything more than a generous and friendly emulation.

o 2

Lady Rodd accompanied the remains of her beloved father to their last resting-place in Westminster Abbey. She then devoted several years to the pious labour of revising new editions of his principal works and of publishing his current charts and memoir. Admiral Sir John Tremayne and Lady Rodd continued to live at 40, Wimpole Street. The admiral died at Tunbridge Wells in 1838. Lady Rodd long survived him, retaining all her faculties to the very last. At the age of seventy-four she met with a severe accident by falling and breaking her thigh—the very same injury which happened to her father. But she recovered, and died, literally of no illness, but old age, on the 14th of December, 1863.

INDEX.

232 INDEX.

PRINTED BY CASSELL & COMPANY, LIMITED, LA BELLE SAUVAGE, LONDON, E.C.

www.ingramcontent.com/pod-product-compliance
Lightning Source LLC
Chambersburg PA
CBHW021658210326
41599CB00013B/1462